A Colour Guide to

CHEESE

and Fermented Milks

A Colour Guide to
CHEESE
and Fermented Milks

RICHARD K. ROBINSON

University of Reading, UK

Editorial advisory board

M. CARIC
University of Novi Sad
Serbia

R.I. FARROW
Gist-brocades
UK

P.F. FOX
University College
Cork, Ireland

F.V. KOSIKOWSKI
Cornell University
USA

CHAPMAN & HALL

London · Glasgow · Weinheim · New York · Tokyo · Melbourne · Madras

Published by Chapman & Hall, 2–6 Boundary Row, London SE1 8HN, UK

Chapman & Hall, 2–6 Boundary Row, London SE1 8HN, UK

Blackie Academic & Professional, Wester Cleddens Road, Bishopbriggs, Glasgow G64 2NZ, UK

Chapman & Hall GmbH, Pappelallee 3, 69469 Weinheim, Germany

Chapman & Hall USA, One Penn Plaza, 41st Floor, New York NY 10119, USA

Chapman & Hall Japan, ITP-Japan, Kyowa Building, 3F, 2-2-1 Hirakawacho, Chiyoda-ku, Tokyo 102, Japan

Chapman & Hall Australia, Thomas Nelson Australia, 102 Dodds Street, South Melbourne, Victoria 3205, Australia

Chapman & Hall India, R. Seshadri, 32 Second Main Road, CIT East, Madras 600 035, India

First edition 1995

© 1995 Chapman & Hall

Typeset in Sabon 10/12½ pt by Keyset Composition, Colchester, Essex

Printed in Hong Kong

ISBN 0 412 39420 0

A catalogue record for this book is available from the British Library

Library of Congress Catalog Card Number: 94-071203

Contents

Editorial advisory board

Preface

Various types of cheese and fermented milk have been produced ever since humans first began to domesticate animals and, as a result, there are now several hundred varieties of cheese manufactured around the world. Some varieties, such as Cheddar cheese, have acquired international renown and are produced in almost every country with access to tonnage quantities of cow's milk. Other equally well-known cheeses – Roquefort is a case in point – are manufactured in limited volumes and under tight regulatory control. Yet despite the universal interest in cheese as a means of preserving milk and/or enriching the diet, published information concerning the production and characteristics of many varieties is extremely sparse.

It was against this background that the idea for an illustrated 'Guide to Cheese and Fermented Milks' came into being, and the original intention was to provide a truly extensive coverage. In the event, locating colour photographs of high technical quality has not been easy and, despite the enthusiastic co-operation of the individuals and organizations cited in the Acknowledgements section, certain varieties have had to be omitted. In other cases, the absence of accessible technical information has proved to be the 'limiting factor'.

Nevertheless, many cheeses that are readily available in shops and supermarkets have been included, and it is to be hoped that potential readers, whether food scientists, technologists, students of dairy science or simply consumers interested in dairy products, will enjoy knowing a little more about the items on their 'cheeseboards'.

R.K. Robinson
University of Reading

Acknowledgements

The authors would like to thank all those individuals and organizations who have so generously provided illustrations of, and/or information about, the cheeses and fermented milks described in this book, and are especially grateful to the following:

Professor S.A. Abou-Donia, Alexandria University, Egypt

Alfa Laval, Lund, Sweden

Professor E.M. Anifantakis, Agricultural College of Athens, Greece

Bayham Foods Ltd, Sevenoaks, Kent, UK

Leo Bertozzi, Consorzio del Formaggio 'Parmigiano Reggiano', Reggio Emilia, Italy

Professor V. Bottazzi, University of Piacenza, Italy

Central Marketing Organisation of German Agricultural Industries (CMA), Bonn, Germany

Centre Interprofessionnel de Documentation et d'Information Laitières (CIDIL), Paris, France

Chr. Hansen's Laboratory Ltd, Reading, UK

Dairy Research Institute, Zilina

Danish Dairy Board, Aarhus, Denmark

Dutch Dairy Bureau, Leatherhead, Surrey, UK

Eden Vale, Middlesex, UK

Farmhouse Cheese Bureau, Wells, Somerset, UK

Food and Wine from France, London, UK

Gregory Ellis Martin & Partners Ltd, London, UK

Hill and Knowlton (UK) Ltd, London, UK

Instituto nazionale per il Commercio Estero, Rome, Italy

Meggle GmbH, Wasserburg, Germany

Ministerio de Agricultura Pesca y Alimentacion, Madrid, Spain

Dr J. Motar, CV INZA, Belgium

National Dairy Council, London, UK

Nestlé SA, Vevey, Switzerland

Norwegian Dairies Association, Oslo, Norway

Dr A. Olano, Instituto Fermentacione Industriales, Madrid, Spain

On-line Instrumentation Inc., New York, USA

Photographic Service, University of Reading, UK

Dr M. Ramos, Instituto Fermentacione Industriales, Madrid, Spain

Ms Maria Sousa, University College, Cork, Ireland

Stork Amsterdam International Ltd, London, UK

Texel (Group Rhône-Poulenc), Épernon, France

Dr A.Y. Tamime, West of Scotland College, Scotland

Tim's Dairy, London, UK

Dr I. Toufeili, American University of Beirut, Lebanon

Ir. M. van Belleghem, Office National du Lait et des Dérivés, Belgium

Wisconsin Milk Marketing Board, Wisconsin, USA

Dr Carlos Zalazar, Instituto de Lactologia Industrial, Argentina.

Finally, the authors would like to place on record their appreciation of the commitment to the project shown by the publishers: the assistance of Nigel Balmforth and the editorial staff at Chapman & Hall has been invaluable.

1 *Cheese and fermented milks – background to manufacture*

Although crude forms of cheese were probably derived almost as soon as humans domesticated animals, the earliest records of cheesemaking are probably those from Ancient Egypt (Abou-Donia, 1991). Earthenware pots of the type still in use today to separate the milk for making the soft Egyptian cheese, Mish, were found in the tomb of King Horaha dating from around 3200 BC, while both reed mats – employed to drain whey from the soft curd – and storage jars for finished cheese have been located in tombs built during the Roman occupation of the Nile basin.

Fermented milks, like yoghurt with its origins in the Middle East and kefir from the Caucasus, probably appeared around the same time, for the practice of storing milk in animal-skin bags would soon have encouraged the development of a microflora capable of fermenting the milk along a fairly predictable pathway. As communities developed a taste for the product in question, so it became an established part of the diet, and the techniques for production became similarly routine. It was likely also that some fermented milks became closer in nature to soft cheese than the typical semi-fluid consistency of present-day products, and certainly the popular Middle Eastern product, labneh, is not unlike quarg or cream cheese in many respects.

By 332 BC, a more textured type of cheese, Domiati, had become established and it is of interest that, like Mish, it is a cheese ideally suited to the conditions pertaining in the Middle East. Thus, the high ambient temperatures and poor standards of hygiene meant that products with high acidities and salt contents were the most amenable to preservation, and it is probable that many varieties popular with present-day consumers, such as Feta or Halloumi, are little more than refinements of these early products.

Variants of these primitive systems of production evolved with time, and it is probable that most tribes with herds of cows, sheep or goats produced some form of cheese or fermented milk. As these groups intermingled as the result of wars or casual migrations, so the art of cheesemaking spread. Climatic or other conditions would have imposed constraints in various ways and, as a result, distinctive local varieties began to emerge.

The introduction of agents capable of rapid coagulation of the milk, such as extracts from plants like the thistle (*Carduus* spp.), fig (*Ficus carica*) or thyme (*Thymus* spp.) would have offered one route towards some degree of standardization of the making process and/or finished cheese, but the role of the bacterial flora could have been equally important.

Clearly there is no evidence of any intentional selection of specific cultures for specific products, but there does appear to be a distinct relationship between geographical location and the types of bacteria associated with fermented products made in the region concerned (Marshall, 1987). Thus, in the colder regions of Europe, many products evolved based upon so-called mesophilic cultures, while in the Middle East and around the Mediterranean, the dominant bacteria tended to be thermophilic, i.e. 'warmth-loving'. This division is not unreasonable, but it does serve to emphasize the potential impact of the environment on the nature of local varieties of cheese or fermented milk.

The broad association of white-brined cheeses with warm climates has been mentioned already, but it is of interest that more localized links between product, culture and environment are also apparent. The derivation of cheeses internally ripened with moulds, e.g. Roquefort or Stilton, is an appropriate example, and certainly with Roquefort the presence of the natural caves for the storage of locally-made cheeses was purely fortuitous. The fact that some of the cheeses developed fissures in the curd into which spores of an appropriate fungus, *Penicillium roqueforti*, became lodged was again pure chance, as was the fact that the environment of the caves provided the ideal balance between aeration and humidity for the development of the mould. Obviously, the local artisans deserve every credit for capitalizing on their discovery but, even so, little could have been achieved but for the 'freak of nature'.

Similar interactions between humans and their environment had resulted, by the Middle Ages, in ranges of cheeses with quite distinctive characteristics, and many of the names referred to in Table 1.1 are still in use today (Scott, 1986; Tamime, 1993). Increasingly, however, human activity played a dominant role in the expansion of cheese production, and changes in human

life-style were, perhaps, the most important of all.

Thus, with the growth of large conurbations, so a section of the consuming public became divorced from production areas, and soft cheeses – well suited for local sale – began, at least in some countries, to lose their market. The reaction of the cheesemakers was predictable enough, for if consumers could no longer reach village markets, then the produce had to be brought into the towns, on transport that was slow and unreliable. As a consequence, harder, drier cheeses became the norm, for it did not take producers long to realize that even if a cheese like Cheddar took three weeks to reach London from Somerset, its quality would be unimpaired, whereas a soft cheese would be rendered totally inedible – even in winter. This pattern became noticeable throughout Europe, and although some countries retained a major interest in soft cheeses, the firmer varieties like Cheddar, Emmental and Gouda became the major varieties of commerce. The advent of refrigeration was to change this pattern decisively, and in the USA, for example, soft cheeses like Cottage cheese returned to claim a major share of the market.

Since those early days, numerous varieties have emerged to attain differing levels of status in the eyes of the consumer but, even so, the essential nature of cheese has changed very little. It is still a product made by the coagulation of milk by enzymes and/or acid with some of the whey expressed from the finished curd, and this simple definition is broadly applicable to all varieties. Different conditions of processing and maturation can dramatically change the nature of the basic curd and, as a result, it is possible to identify a number of basic types of cheese. These groupings are summarized in Table 1.2.

However, in spite of these apparently major divisions between the types of cheese, and the fact that each division may contain many, quite distinct varieties, the fundamental features of cheese remain immutable. Most fundamental of all is the raw material, milk, and before discussing the actual process of cheesemaking it is appropriate to consider the essential characteristics of this basic ingredient.

Table 1.1 Some of the cheese names for which early records exist

Name	Date (AD)	Name	Date (AD)
Gorgonzola	879	Roquefort	1070
Cheshire	1085	Maroilles	1174
Schwangenkäse	1178	Grana	1200
Taleggio	1282	Gruyère	1288
Cheddar	1500	Parmesan	1579
Emmental	1622	Dunlop	1688
Gouda	1697	Gloucester	1783
Stilton	1785	Camembert	1791
Limburg	1800	St Paulin	1816

After Scott (1986); Tamime (1993).

Table 1.2 Some major types of cheese that can be recognized on the basis of moisture content and/or method of maturation

Description	Popular examples	Country
Extra-hard cheese (grating)	Grana Padano	Italy
	Parmigiano (Parmesan)	Italy
	Passelan	France
Hard cheeses	Cantal	France
	Cheddar	England
	Leicester	England
	Manchego	Spain
Semi-hard cheeses	Caerphilly	Wales
	Port du Salut	France
Cheeses with 'eyes'	Edam	Holland
	Emmental	Switzerland
	Gruyère	Switzerland
	Maribo	Denmark
Cheeses internally ripened with moulds	Danablu	Denmark
	Gorgonzola	Italy
	Roquefort	France
	Stilton	England
Cheeses surface-ripened with moulds	Brie	France
	Camembert	France
Cheeses surface-ripened with bacteria	Bel Paese	Italy
	Klosterkäse	Germany
	Limburger	Belgium
	Tilsit	Germany
Unripened soft cheeses	Cottage cheese	USA
	Cream cheese	USA
	Fromage frais	France
Italian-style cheeses (pasta filata)	Cotronese	Italy
	Mozzarella	Italy
	Provolone	Italy
White-brined cheeses	Feta	Greece

Milk for cheesemaking

The milk from any mammal can, in theory, be turned into a cheese-like product but, for purely practical reasons, milk from domesticated animals has always dominated production. The primary source has tended to be determined by the vagaries of the local countryside, so that while cow's milk has been more readily available in lowland areas, mountain tribes have relied on sheep or goats as the sources of raw material. Subtle differences between the various milks, such as levels and composition of the milk fat, do affect the nature and quality of any cheese produced but, in many respects, the milks have much in common. The overall chemical compositions of the important milks are shown in Table 1.3.

The two major forms of protein that occur in milk (namely the caseins and the whey proteins) are essential for cheesemaking and of these the caseins are the most important. Caseins not only represent around 70–80% of the total protein present in milk, but also they are the proteins which form the matrix of the curd and, ultimately, the cheese itself. Only if the cheese is made from milk that has been subject to ultra-filtration or severe heat-treatment is the whey protein fraction of real importance; in the former case, the whey proteins are trapped in the process milk, while in the latter they will be denatured and become adsorbed onto the surface of the casein.

The casein fraction consists of four types of casein – α_{s1}, α_{s2}, β and κ – and each one is a distinct phosphoprotein. The presence of phosphorylated residues in each protein is important in that (a) it makes the units acidic in nature with isoelectric (coagulation) points at around pH 4.5–5.0, and (b) the phosphorylated residues can bind to calcium ions.

The molecules of casein do not exist in milk in isolation, but combine with calcium and calcium phosphate to form so-called casein micelles. These complex units contain several thousand casein molecules. The centre of the complex is composed largely of α- and β-casein but much of the surface consists of κ-casein (Dalgleish *et al.*, 1989). It is the presence of the κ-casein towards the outside of the micelles that prevents the α- and β-caseins from aggregating and precipitating spontaneously.

However, when the milk is coagulated during cheesemaking, the enzyme chymosin, present in rennet or some similar preparation, splits the

Table 1.3 Chemical compositions of milks from the important species of mammal (g/100 g liquid milk)

Type	Water	Fat	Protein	Lactose	Ash	Calcium
Buffalo	82.1	8.0	4.2	4.9	0.8	
Camel	87.1	4.2	3.7	4.1	0.9	
Cow	87.6	3.8	3.3	4.7	0.6	0.08
Goat	87.0	4.5	3.3	4.6	0.6	0.95
Mare	89.0	1.5	2.6	6.2	0.7	
Sheep	81.6	7.5	5.6	4.4	0.9	

After Tamime and Robinson (1985).

κ-casein, so leaving the underlying caseins unprotected. As a result, the micelles become prone, in the presence of calcium ions, to aggregate together to form a curd. The effect of the enzyme on the other caseins is negligible in the short term, but in cheeses with long maturation times the α- and β-caseins may be attacked by the proteolytic enzymes (e.g. pepsin) which are normally present in rennet along with the chymosin. This action can be extremely important in relation to flavour development (Fox, 1989), for as different species of mammal have slightly different forms of casein, the nature of any developed flavours may vary as well. In cow's milk, for example, the α_{s1} and β-caseins are the principal components, but in sheep's and goat's milk two forms of β-casein are the important fractions, and both forms are different from the β-casein in cow's milk. Similarly, α_{s2} is at a higher level in sheep's or goat's milk than in cow's milk, and subtle differences like these could well be important in respect to the ultimate flavour profile of a given cheese.

When acid alone is employed for coagulation of milk, as is practised in India or South America, for cheesemaking and world-wide for the manufacture of products like yoghurt, the κ-casein remains intact, and it is the instability of the caseins around their isoelectric point that is exploited. The result is a soft gel which is then manipulated in a variety of ways to produce specific organoleptic properties in the retail product. Although all milks can be 'gelled' in this fashion, the nature of the caseins is important

with respect to gel strength; the β-caseins that dominate goat's milk, for example, give rise to a very weak gel in comparison with that formed in cow's milk.

The other major component of most cheeses is the lipid fraction, and some typical values for the fat content of milk might be: cow's milk 3.5–5.0%, sheep's milk 5.8–9.0% and goat's milk 2.8–6.5% (Tamime *et al.*, 1991). However, such figures are so prone to variation as a consequence of breed or diet that the only important generalization is that sheep's milk usually has the highest fat content.

In general, milk fat is present as globules surrounded by a membrane (fat globule membrane) composed mainly of phospholipids. The maximum size of the globules is around twelve microns, but in sheep's milk only around 2.5% have a diameter above six microns; in goat's milk the fraction is slightly higher, but only in cow's milk does the level of globules having large diameters reach double figures (>15%). It has been suggested that these large fat globules tend to give cow's milk its slightly yellow appearance, and certainly cheeses made from sheep's milk tend to be whiter than their bovine counterparts.

The free lipid within the membrane is mainly composed of triglycerides, although free fatty acids are widely present in the body of the milk. The saturated nature of the fatty acids in milk fat has raised speculation about the health implications of consuming dairy products but, in spite of considerable publicity, the evidence appears to be open to a variety of interpretations.

During cheesemaking, the lipid content of the milk is retained in the curd along with the protein, so that in a typical hard-pressed cheese a minimum of 45% of the solids content is fat; the remaining solids being protein, along with low levels of lactose and mineral salts.

Although most of the lactose is lost into the whey fraction during the manufacture of cheese, this disaccharide sugar – composed of one molecule of glucose and one of galactose – is essential for the early stages of cheesemaking, for it is this sugar that is so readily metabolized by the starter bacteria to produce lactic acid. The extent of this acidification stage depends on many factors, including the types of bacteria present and the variety of cheese being manufactured, but the essence is that:

- the acid conditions are necessary for the coagulating enzymes to work effectively;
- adventitious bacteria that could lead to the development of off-flavours in the maturing cheese are suppressed; and
- the lactic acid contributes to the flavour of many varieties.

The lactose also supports, at least in part, the growth of those moulds, yeasts and bacteria associated with the ripening of varieties like Stilton or Brie, as well as the gas-forming bacteria that give rise to the 'eyes' in so many cheeses from mainland Europe. Consequently, although little lactose is to be found in most varieties of cheese, it is a vital component of the raw material, as are the mineral salts.

Some minerals, like calcium and phosphorus, are essential for the process of milk coagulation, and calcium ions in particular play a central role in the aggregation of the casein micelles. This retention in the curd is equally valuable from a nutritional standpoint, in that it makes cheese an exceptional source of dietary calcium.

In fermented milks from which no whey is expressed, all the residual lactose, together with the minerals, is retained and this fact has potential implications for the consumer. In particular, the level of lactose in yoghurt and similar products is little different from that in milk but,

whereas milk cannot be tolerated by those communities with individuals lacking the intestinal enzymes necessary to metabolize lactose, the consumption of yoghurt produces no real discomfort. The main reason for this effect appears to be that the bacterial cultures, and/or their enzymes, continue to act long after the cultured milk has been eaten and so, in effect, make up for any inherent deficiency in the consumer. In addition, some of the bacteria associated with fermented milks do become implanted in the human intestine, and this beneficial attachment again ensures that any lactose entering the system is metabolized before any adverse symptoms are recorded (Robinson, 1991; Salminen and von Wright, 1993).

The retention of the mineral salts is also relevant, because the acid environment of cultured milks tends to increase the bio-availability, for example, of calcium and zinc, both of which are well represented in milk and essential for human metabolism. The availability of other ions could be equally important, and a detailed survey of the nutritional properties of yoghurt has been published by Tamime and Deeth (1980); much of the data is, in general, applicable to similar products.

Starter cultures for cheese and fermented milks

Bacterial cultures

A starter culture can be defined as an agent employed to bring about a fermentation process, and consisting of one or more species of bacterium – and usually two or more strains of these same species, together with an inert carrier such as lactose. The bacteria are selected for their ability to produce lactic acid from lactose and/or their ability to produce distinctive metabolites that play a role in the derivation of the flavour profile of a given cheese or fermented milk. In the latter context, the bacteria are especially important in cheeses that undergo a long period of maturation, because:

- Their metabolic activities during the early stages of maturation create the reducing conditions necessary for the numerous biochemical changes associated with flavour development to occur.
- The release of various enzymes from dead cells again contributes to the complex changes that turn the raw curd into the finished cheese.

The following genera of bacteria are important in dairy fermentations.

Streptococci

Some of the best-known members of this genus have now been transferred to the genus *Lactococcus* (see below), but one important subspecies that has not been re-classified is *Streptococcus salivarius* subsp. *thermophilus*. This organism – more frequently cited as *Str. thermophilus* – is an essential component of the cultures employed to manufacture yoghurt. It is a Gram-positive coccus that appears as long chains when growing in milk (Figure 1.1), and at its optimum temperature of around 37 °C it rapidly ferments lactose to lactic acid.

Leuconostocs

This genus of Gram-positive cocci is frequently employed in dairy cultures, often at low levels, to provide additional flavour compounds. Thus, unlike many cultures that are homofermentative (i.e. produce principally lactic acid when growing in milk), leuconostocs produce carbon dioxide and alcohol as well – heterofermentative metabolism. Acetic acid is secreted also as a product of citrate breakdown, so that when employed in a cheese or fermented milk alongside *Lactococcus lactis* biovar *diacetylactis* (see below), the flavour profile of the product can prove most attractive.

The important species is *Leu. mesenteroides* subsp. *cremoris* – formerly known as *Leu. cremoris* – and it is available in many commercial starter cultures at a level of 5–10% of the total bacterial population; acid-producing lactococci will tend to be the dominant organisms present.

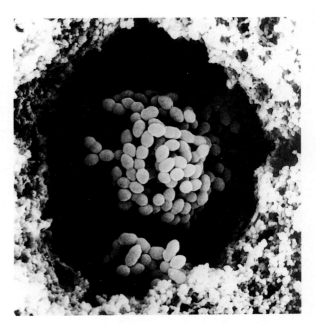

Figure 1.1 A typical colony of *Streptococcus salivarius* subsp. *thermophilus* as it appears in natural yoghurt; note the growth of the bacterium in discrete chains. (Microphoto SEM by Bottazzi-Bianchi, Institute of Microbiology, UCSC, Piacenza, Italy; with permission of Elsevier Science Publishers Ltd, UK.)

Lactococci

Members of this genus (previously placed in the genus *Streptococcus*) are all Gram-positive cocci, and are characterized by having an optimum temperature for growth of 20–22 °C. A higher temperature (30 °C) is often used during cheese-making in order to reduce the fermentation time, but with cultured (mesophilic) milks, the lower range tends to produce a better quality end-product. The important species of this genus are:

- *Lactococcus lactis* subsp. *lactis*
- *Lactococcus lactis* subsp. *cremoris*
- *Lactococcus lactis* biovar *diacetylactis*

As shown in Table 1.4, they are employed in the manufacture of a wide range of cheeses. The two subspecies of *Lact. lactis* are homofermentative (i.e. release few metabolites into the milk except lactic acid) and hence are used principally to acidify the cheesemilk during the ripening stage, and then continue to produce lactic acid until the

Table 1.4 Some of the cheeses manufactured with the different types of starter culture shown (the precise combination of bacteria depends upon the variety of cheese and the inclination of the cheesemaker)

Micro-organisms	Varieties of cheese
Mesophilic group	
Lactococcus lactis	Cheddar, Cheshire, Caerphilly, Edam,
Lact. lactis subsp. *cremoris*	Gouda, Camembert, Brie, Danish Blue,
Lact. lactis biovar *diacetylactis*	Stilton, Cottage and Cream cheese
Leuconostoc cremoris	
Propionibacterium spp.	Emmental, Gruyère
Thermophilic group	
Streptococcus salivarius subsp. *thermophilus*	
Lactobacillus delbrueckii subsp. *bulgaricus*	Parmesan, Romano, Emmental, Gruyère
Lac. helveticus	
Lac. lactis	
Moulds	
Penicillium camemberti	Camembert, Brie
Penicillium roqueforti	Danish Blue, Roquefort, Stilton

acidity desired in the finished cheese has been achieved.

The biovar *diacetylactis* is similar to *Lact. lactis* subsp. *lactis* except for the ability to produce diacetyl from citrate. As this feature is plasmid controlled, and hence easily lost during routine multiplication of the organism, it is given the status of biovar rather than subspecies. However, this heterofermentative ability of *Lact. lactis* biovar *diacetylactis* is important, for its major role is to provide a more complex flavour in the finished cheese. Trace amounts of carbon dioxide are also liberated by heterofermentors, and the presence of such organisms may tend to open the texture, albeit modestly, of an otherwise 'close' cheese like Cheddar.

The same three organisms are also widely used for the production of fermented milks in Scandinavia, for example filmjolk, and the slightly 'buttery' flavour associated with diacetyl has proved widely popular in these northerly regions. An interesting variant of the subspecies *lactis* is employed in some brands, and it is one that produces an extracellular protein that gives the retail item a distinctive, viscous or slimy mouthfeel. It is unusual in that most bacteria associated with fermented foods build up polysaccharides as

an external coating; the subspecies does not appear to figure in fermentations outside Scandinavia. Elsewhere, the use of these mesophilic cultures is largely restricted to the production of buttermilk and cultured cream in North America, or leben in North Africa.

Although the three organisms are clearly identifiable at subspecies/biovar level, numerous variants or strains of these species abound. Differences between strains are usually subtle, and certainly not distinctive enough to characterize any of the variants as biovars. Nevertheless, the rate of acid production is one feature that can vary with strain, and one that is important to the cheesemaker, e.g. the starter for Cheshire cheese needs a more active rate of acidification than a culture for Cheddar cheese. Equally important is the reaction of different strains to viruses, or bacteriophages (phages) as they are more commonly known. The size of these phages is usually around 150–200 nm, and they are composed of a head region containing the genetic material, usually DNA, and a protein tail. In commercial practice, the important phages are the so-called virulent types, for these enter the cell of a susceptible bacterium, and then rapidly replicate to form new phages. Up to 200 new virus particles can be formed within one bacterial cell before it finally bursts, releasing new phages into the medium (Cogan and Accolas, 1990). The end-result is that, if the starter culture is susceptible to a phage-type present in the cheesemilk, the number of bacterial cells can be drastically reduced, and the essential production of lactic acid may be prevented. The total loss of several thousand gallons of milk may be the outcome.

In theory, it is possible to select strains of starter bacteria that are resistant to all phages present in a given location, and some dairies in New Zealand have isolated cultures for the manufacture of Cheddar cheese which can be used on a year-round basis without loss of activity through infection. However, most dairies rely on a rotation of cultures on a daily or weekly basis (depending upon the system employed), so that the levels of phage active against a specific strain of bacterium are never allowed to build up to sufficiently high levels to cause problems.

Lactobacilli

While many hard and semi-hard varieties rely on cultures that include species of *Lactococcus*, the varieties of Swiss cheese, amongst others, employ cultures in which species of *Lactobacillus* dominate, and the same genus is central to the production of many fermented milks.

This latter genus includes a number of species that grow well in milk as short, Gram-positive rods. This structure is clearly visible in Figure 1.2, and the organism illustrated (*Lac. delbrueckii* subsp. *bulgaricus*) is often used, along with *Str. thermophilus*, in the manufacture of Italian cheeses. Other species of lactobacilli which are important are *Lac. helveticus*, *Lac. casei* subsp. *casei* and *Lac. plantarum*; all these species are characterized by their optimum growth temperatures of 37–45 °C, and the fact that they are homofermentative. This restriction means that their role is essentially as 'acid-producers', and the rich flavour of Italian cheeses, for example, arises from enzymes added to the cheese coupled with a long maturation time, rather than direct bacterial activity.

Aside from its presence in Italian cheese cultures on the grounds of its thermotolerance, *Lac. delbrueckii* subsp. *bulgaricus* (*Lac. bulgaricus*) is widely employed along with *Str. thermophilus* in the production of yoghurt. This universal application of the two subspecies arises because of their unique physiological interaction: whilst *Str. thermophilus* secretes lactic acid, carbon dioxide and formic acid, all of which stimulate the growth of *Lac. bulgaricus*, the latter breaks down some of the milk protein required for growth of the streptococci. The end result is a rapid fermentation of the milk to give yoghurt in around four hours, whereas the same cell count of either organism growing alone would require many hours longer to liberate the same level of acidity.

Although it grows only very slowly in milk, another species with a long association with fermented dairy products is *Lac. acidophilus*. This species, which is a natural inhabitant of the

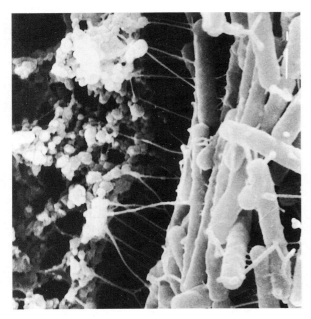

Figure 1.2 Typical cells of *Lactobacillus delbrueckii* subsp. *bulgaricus* as seen in natural yoghurt; the rod-like structure is common to many lactobacilli, but the production of extracellular polysaccharides – seen as thin strands connecting the bacteria to the protein of the yoghurt – is only observed in certain species. (Microphoto SEM by Bottazzi-Bianchi, Institute of Microbiology, UCSC, Piacenza, Italy; with permission of Elsevier Science Publishers Ltd, UK.)

intestinal flora of human beings, was probably an adventitious contaminant of many artisanal starter cultures, but more recently it has been exploited in its own right. Originally, it appeared on the market in North America in Acidophilus milk, a mildly acidified milk-drink that was consumed mainly for its alleged health-benefits (Sellars, 1991) but nowadays Sweet Acidophilus milk or 'Bio-yoghurts' are its principal consumer vehicles. Sweet Acidophilus milk is normal pasteurized milk to which a culture of *Lac. acidophilus* is added but not allowed to ferment, so that the flavour of the product is unaltered *vis-à-vis* the base material. Bio-yoghurts – known variously as 'mild yoghurts' or 'B/A yoghurts' – have many of the properties of traditional yoghurt, except that *Lac. acidophilus* is employed as a component of the active culture or, alternatively, is incorporated into a base formed by the activity of a standard yoghurt culture. Either way, the essential requirement is that the *Lac. acidophilus* should be:

- of human origin, i.e. it is a strain that is capable of surviving the human digestive process and colonizing the lower end of the small intestine;
- present in the product at a level of around one million cells per ml (otherwise too few reach the lower intestine to have a desirable influence).

In some parts of the world, *Lac. casei* is employed in a similar fashion, and for the same reasons.

Bifidobacterium

The natural habitat of this genus is the large intestine of warm-blooded animals, including humans. They are Gram-positive organisms that vary in morphology from rods to Y-shaped structures, and most strains are intolerant of oxygen. In the large intestine, they perform a number of valuable functions (Kurmann and Rasic, 1991; Robinson and Samona, 1992), and it is for this reason that a number of manufacturers are including bifidobacteria in the cultures for various fermented milks and/or bio-yoghurts.

Rather in the manner of *Lac. acidophilus*, bifidobacteria are poorly adapted to growing in milk, and hence have to be used alongside other cultures. A combination of *Bifidobacterium* sp. and *Lac. acidophilus* is employed for some 'health-promoting' products like the 'Cultura' range found in Denmark (see later), but more usually a low level of yoghurt culture (or *Str. thermophilus* alone) is added as well, both to speed up the fermentation and to enhance the flavour of the end-product.

Propionibacterium

One final species that falls within the definition of starter cultures is *Propionibacterium freudenreichii*, which occurs as three subspecies: *freudenreichii*, *globosum* and *shermanii*. All three are employed in the manufacture of Swiss and other cheeses with 'eyes', for the late metabolism of the propionibacteria releases carbon dioxide that forms holes or cavities in the partially matured curd. The nature of the curd, together with specific features of the bacteria and maturation procedure, determines whether the final cheese will have a few large holes or many small ones but, either way, the role of this genus is critical.

Although present as a constituent of the natural flora of the cheeseroom (as against being introduced as a defined culture), *Brevibacterium linens* is essential for the successful ripening of cheeses like Brick and Limburger. These cheeses, which are usually referred to as 'smear-coated', are distinctive in that the outer surface of the cheese becomes orange-brown in colour and greasy to the touch as a result of the growth of this bacterium; the flavour of the cheese is enhanced as well. The importance of the bacterium with respect to other cheeses is less clear-cut, but it does form a minor part of the surface flora of Camembert, for example, and may well contribute to the overall flavour profile of a number of varieties.

Mould cultures

Although species of *Penicillium* are the dominant organisms associated with mould-ripened cheeses, the final flavour depends also on the presence of yeasts and bacteria growing on the surface of the cheese. This growth may take place prior to the development of the mould or, more usually, concomitantly, and the exact contribution of yeasts to the flavour profile of a cheese like Brie or Camembert is difficult to define.

The moulds themselves are grouped, at least initially, in relation to the colour of the spores, so that the dairy industry recognizes two groups: 'white moulds' and 'blue moulds'. The former group is typified by one species, *Penicillium camemberti*, and various strains are widely employed for the production of Brie and Camembert.

The major impacts of this growth are the liberation of flavour compounds, both by lipolysis of the milk fat and by the synthesis of secondary metabolites by the mould, together with a softening of the curd by proteolytic enzymes released by the fungus. The end result is a soft product that, at full maturity, is clearly spreadable and has a distinctive 'peppery' taste.

The blue moulds are typified by *Penicillium roqueforti*, a species which grows within the curd of cheeses like Danish Blue, Gorgonzola, Mycella, Roquefort and Stilton. Although the mycelium is colourless, exposure to air encourages development of the characteristic blue/blue-green spores, and visible veins appear throughout the cheese. As the mould grows, so it secretes a lipase which hydrolyses the milk fat to produce a range of fatty acids, such as caproic, caprylic and capric acid. It is these acids which give the cheese its slightly 'peppery' flavour, whilst the corresponding ketones, such as methyl-*n*-amyl ketone from caprylic acid, provide the well-defined blue-cheese flavour. The breakdown of milk proteins is important also (Seth and Robinson, 1988).

Physical forms of culture

Traditionally, the essential bacterial flora for cheesemaking would have been derived from the utensils in everyday use or, for fermented milks in particular, a small portion of product from the previous day would have been employed. Gradually, however, this approach was refined so that each dairy set aside special vessels in which bacterial cultures were grown for use on the following day. As factories became larger, so well-defined culture systems evolved. In general, the procedure involves:

- growing the desired culture in the laboratory;
- producing a larger volume in the laboratory or on the factory-floor (the so-called mother culture);
- producing a bulk starter for direct addition to the cheese vat, with the volume being around 2% v/v of the volume of cheesemilk to be inoculated (Figure 1.3).

Whole milk, or more usually skim-milk, provides the growth medium for the various cultures. To avoid contamination by stray bacteria or phages, the milk is severely heat-treated at each stage prior to inoculation.

This approach is still widely employed worldwide, but with the important difference that specialist culture suppliers now provide the basic cultures rather than the dairy having to maintain them 'in-house' (Figure 1.4).

The different bacteria are available as individual species or in combination, e.g. a defined blend of *Lact. lactis* subspp. *lactis* and *cremoris*, either deep-frozen at $-196\,°C$ or as freeze-dried powders, and in volumes suitable for inoculating up to 1000 litres or more of milk. This trend has meant that many dairies are able to dispense with the laboratory stages of culturing altogether, and simply add the 'off-the-shelf' starter direct to the bulk starter vessel. As a result, problems with poorly balanced or inactive cultures have been dramatically reduced, and standardization of this stage of cheesemaking has become comparatively straightforward.

Today, this system has been taken a stage further in that cultures are now available for direct addition to the cheese vat, and the manufacture of much of the cheese sold in industrialized countries is based upon these types of culture. The same is true also of fermented milks

Figure 1.3 After incubation to build up a high level of bacteria, followed by overnight cooling, the bulk starter culture is ready for addition to the cheese milk. (Courtesy of Group Rhône-Poulenc, France.)

(Figure 1.5), and the attractions for the producer can be summarized as:

- maximum convenience;
- reliability of cultures leading to product of consistent quality;
- lower risk of infection of a culture by undesirable bacteria, yeasts or phages;
- financial savings, especially if the dairy is able to avoid the capital cost of installing expensive equipment for starter production.

Coagulants and coagulum formation

Coagulation of the milk is one of the essential steps in cheesemaking. According to Tamime (1993), it can be achieved by:

- acid alone, usually lactic acid generated by the starter culture, but citric acid is employed for some soft cheeses (Chandan, 1991);

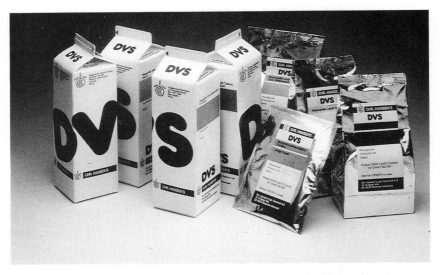

Figure 1.4 A group of direct-vat-set (DVS) cultures of the type that can be employed to manufacture a bulk starter or be added direct to the process milk. The cultures in cartons are deep-frozen pellets (see also Figure 1.5), while the sachets contain similar cultures in freeze-dried form. (Courtesy of Chr. Hansen's Laboratory, USA.)

Figure 1.5 Deep-frozen cultures are widely available for the direct inoculation of the process milk, and most are in the form of frozen granules that melt readily in the milk so that the bacteria are rapidly dispersed. This example contains *Lac. acidophilus* and *Bif. bifidum* for use in a bio-yoghurt or similar product. (Courtesy of Chr. Hansen's Laboratory, USA.)

- the addition of coagulant alone, e.g. Domiati, although some acidity may be derived from the natural microflora of the milk;
- a combination of acidity and coagulant as employed for the majority of cheeses.

In general, therefore, the other essential addition to the process milk for the manufacture of cheeses is an enzyme or enzymes to coagulate the caseins. The following coagulants are most widely employed:

- Enzymes of animal origin extracted from the stomachs of calves or other young animals, known commercially as rennets.
- Microbial enzymes obtained from growing cultures of moulds, such as *Mucor meihei* or *Mucor pusillus*.
- Extracts from certain plants, such as the fig (*Ficus carica*), have sufficient proteolytic activity to coagulate milk and are used locally in Africa and elsewhere.
- Certain micro-organisms, such as *Kluyveromyces marxianus*, have been genetically modified to produce an enzyme, chymosin, which is identical in action to that found in the stomachs of young animals.

All these enzymes are, in effect, proteinases, and the most popular rennet consists of a mixture of chymosin (>75%) and pepsin. Other ratios are possible, but the broad proteolytic activity of pepsin can have an adverse effect on the texture or flavour of some cheeses.

It is for this reason that chymosin, both natural and derived, is the most popular, because the action of the enzyme is quite specific in that it hydrolyses just one bond between the amino acids phenylalanine and methionine (nos. 105–106) in the κ-casein. One product, *para*-κ-casein (residue 1–105), aggregates along with the α- and β-casein to form the coagulum, whilst the other section, the caseinomacropeptide (residue 106–169), is eventually lost into the whey (Tamime *et al.*, 1991). Most of the coagulant is also lost into the whey, but around 10% may be retained in the finished cheese (Fox, 1989). Pepsin, in particular, may play an important role in the maturation process, for the partial proteolysis that occurs contributes to both flavour and texture. However, in fresh cheeses — short shelf-life, or in Swiss cheeses which are subject to heating above 50 °C during manufacture — the coagulant will be inactivated and the retention of the enzyme is unimportant.

The cheesemaking process

Initial handling

All milk for cheesemaking was, at one time, collected in churns (Figure 1.6), and the handling of small volumes of milk is still essential in many parts of the world (Figure 1.7). However, in the major dairy countries the cooled raw milk — usually cow's milk — is now transported to the dairy in insulated tankers (Figure 1.8). The method of initial handling does, of course, affect the microbiological quality of the milk, and high bacterial counts can have a detrimental impact on the quality of the cheese. Those groups of bacteria that can grow at refrigeration temperature — the psychrotrophic flora — are of most concern in industrialized countries, because the proteolytic and lipolytic enzymes secreted by these organisms

Figure 1.6 In the early days of cheesemaking, all milk would have been collected in churns, and most farmers built roadside platforms so that the churns would be at the same height as the body of the collection vehicle.

can influence both coagulum formation and the flavour of the finished cheese. For example, Law *et al.* (1976) found that Cheddar cheese made from milk with a high count of psychrotrophs was rancid in taste after around four months, and hence methods to avoid this potential problem (e.g. on-farm pasteurization of the milk and/or cooling to 2 °C) have become widespread.

Another major microbiological problem can arise from the presence of spore-forming bacteria, such as *Bacillus* or *Clostridium* spp., for both genera are frequent contaminants of milk from soil, grass and elsewhere, and their spores withstand normal pasteurization. Clostridia, in particular, have been associated with problems of

Figure 1.7 Centres for the collection and cooling of small volumes of milk are still important in rural areas of India and elsewhere. (Courtesy of Nestlé SA, Switzerland.)

'late blowing' in cheeses, such as Emmental and Gouda, when spores germinate during maturation and liberate carbon dioxide; the development of a rancid flavour usually accompanies such growth. The removal of excessive levels of spores from milk is not easy, for whilst cloth filters are employed almost universally to extract gross contaminants like straw or hairs, few spores are collected in the process. Consequently, additional processes, such as centrifugation (bactofugation), microfiltration or the use of inhibitors (Tamime, 1993) have to be included in the sequence of cheesemaking operations where bacterial spores are considered a risk.

However, while problems with spores tend to be associated with certain types of cheese, difficulties arising from variations in the chemical composition of milk resulting from changes in feed or season of the year and/or differences

between breeds have to be monitored with care. This aspect is of especial importance for industrial-scale operations, as seasonal variations can influence both the yield of cheese obtained and its quality. Some typical variations are shown in Table 1.5 and, bearing in mind that the desired

Table 1.5 Some national averages (UK) for milk fat (%) and protein (%) during 1980/1

Month	Milk fat		Protein	
	Friesian	Jersey	Friesian	Jersey
January	3.98	5.53	3.24	3.90
March	3.93	5.37	3.18	3.78
May	3.71	5.03	3.37	3.89
July	3.80	5.11	3.42	3.97
September	3.94	5.32	3.54	4.08
November	4.07	5.64	3.41	4.11

After Tamine and Robinson (1985).

Figure 1.8 In most industrialized countries, raw or on-farm pasteurized milk is transported to creameries in bulk tankers. After quality checks on the milk in the tanker (e.g. temperature of the milk and microbiological quality), the milk will be stored in silos until used.

casein to fat ratio for the manufacture of Cheddar cheese is 0.68–0.72 (Tamime, 1993), the need for standardization is self-evident. Automated techniques for monitoring and adjusting fat:protein ratio are widely available (Tamime and Kirkegaard, 1991), and a typical monitoring system is shown in Figure 1.9.

Testing the milk for inhibitory substances, such as detergent or disinfectant residues or antibiotics used to treat the animal itself, is essential in large creameries, for the usual starter cultures are extremely sensitive to such materials. Levels of penicillin in the region of 0.02 IU/ml of milk will cause severe retardation of starter activity, and with modern cheese vats holding several thousand gallons, failure of acid development can be serious indeed.

The other important chemical considerations at this stage relate to calcium levels and colour. Calcium ions are essential for successful coagulation of the milk, and whilst the levels indicated in Table 1.3 are usually adequate, the addition of calcium chloride may be necessary in some parts of the world, or with milks other than cow's milk. Similarly the natural colouring matter in milk (usually carotenes derived from the feed) may provide the colour desired in the finished

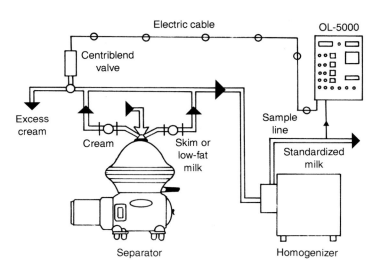

Figure 1.9 A typical system for the automatic standardization of the fat content of milk. Samples taken after the homogenizer are assessed for fat content; any variations from a set value are rectified by the centriblend valve. (Courtesy of On-Line Instrumentation Inc., New York, USA.)

Figure 1.10 Side view of popular design of plate heat exchanger consisting of a stand upon which are suspended a series of plates with caulking strips between. When clamped together as shown, milk passes as a thin film up and down one side of the plate and hot water up and down the other. Consequently, the milk heats very quickly; and rapid cooling can be achieved by replacing hot water with chilled water in the following section.

cheese, but for more distinctively coloured cheeses, such as Cheshire or Red Leicester, annatto has to be added. This water-soluble dye is extracted from the fruits of the bush *Bixa orellana*, and by varying the rate of addition to the cheese milk products from light yellow through to deep orange–red can be obtained. By contrast, excess colour derived from the milk is often a problem with white cheeses like Feta, particularly if cow's milk is used as against the traditional, less coloured, sheep's milk. As removal or destruction of the residual colouring material is impractical, chlorophyll-based dyes are sometimes added in order to mask the natural yellow hues of the cow's milk, but this practice is probably not widespread.

Production of the curd

Once any modifications to the base milk have been made, the milk is normally pasteurized at 72 °C for 15 seconds prior to cooling to 30 °C and transfer to the cheese vat. This high-temperature–short-time (HTST) process (Figure 1.10) is in almost universal use in the major dairy countries, but many small-scale producers, or those that insist on traditional technology, may still use raw milk. Whether or not the presence of the natural microflora has any real influence on the flavour of a cheese is a moot point, but true connoisseurs of a particular variety always maintain that heat treatment has an undesirable effect. For most consumers, however, it is likely that any difference would be imperceptible, while the risks of contracting a food-borne disease are real enough. Obviously with a long maturing cheese, the risk of a milk-borne pathogen remaining viable is reduced, but the growing interest in soft cheeses makes the acceptance of pasteurization a vital aspect of consumer safety.

A wide variety of vats are employed throughout the industry, with the final choice being based upon the type of cheese to be produced and the scale of the operation. Traditionally, square open vats of the type shown in Figure 1.11 were in common use and, except for the paddle-type agitator employed for mixing the rennet and starter culture into the milk and stirring the cut curd, all the operations were performed manually. Later, round-ended, totally enclosed, stainless steel vats of several hundred to several thousand litres became the 'norm' (Figure 1.12), and many variations on this design are in use today (Tamime, 1993). Copper tanks are still used in small factories making cheeses like Emmental or Italian cheeses of the Parmesan type, but otherwise stainless steel enjoys universal application.

Figure 1.11 An example of an early cheeseroom dedicated to the manufacture of hard-pressed cheeses. By removing the wedges, the tanks could be tilted to facilitate drainage of the whey, and circulation of water through the jacket of the vessel controlled the temperature of the milk.

Figure 1.12 Much cheese is still manufactured world-wide in round-ended vats. The cutters/stirrers move along the length of the vat during operation, and the direction of rotation of the cutters is often reversible: a knife-edge is used for cutting, while a blunt 'rear' edge becomes the 'front' during stirring.

Once in the vat, the starter culture will be added (Figure 1.3) and, after an initial period to allow bacterial acidification of the milk, the coagulant will be added. Some 25 ml/100 litres of milk represents a typical rate of addition. To ensure effective mixing of the rennet throughout the milk, it is normal practice to dilute the enzyme with 3–4 times its own volume of water immediately prior to addition.

Handling of the curd: hard-pressed cheeses

After the milk has set to give a firm coagulum (the exact time will depend on a number of factors such as starter activity, temperature, type of milk, etc.) the gel is cut. Rotating stainless steel knives – arranged vertically and horizontally about a central shaft (Figure 1.12) – are used to cut the gel into cubes of the desired size. The precise dimensions are adjusted both for the type of cheese (3–4 mm for Parmesan; 6–8 mm for Cheddar; 12 mm for Edam or Gouda) and the wishes of the cheesemaker. As the cutters revolve at slow speed, it may take about 15 minutes for the coagulum to be reduced to a mass of curd particles floating in whey, at which point heating of the curd/whey mixture can begin. In the manufacture of some Dutch and Swedish cheeses, about one-third of the whey is removed at this stage and replaced with warm water because, by removing some of the residual lactose, less acid is produced by the starter bacteria and a more mild cheese can be produced.

This stage of heating, or scalding as it is usually

known, is common for most cheeses, and involves raising the temperature of the curd/whey mixture by around one degree per minute to a preset level determined by the nature of the cheese. For example, some so-called 'high scald' cheeses like Emmental are heated to 52–54 °C, while the low to mid-30s is usual for most hard and semi-hard cheeses, such as Cheddar, Cheshire or Wensleydale. The effect of this treatment has been summarized by Tamime (1993) as follows:

- The casein network alters to give a semi-rigid structure to the particles, together with a firm 'coat' around each discrete piece of curd that reduces losses of fat into the whey.
- The application of heat to this casein network causes the particles to shrink and expel some of the trapped whey.
- The starter bacteria embedded in the curd pieces continue to generate lactic acid, an action that further encourages expulsion of the whey. However, at high scald temperatures (e.g. around 50 °C), the curd tends to take on a plastic texture which acts to restrict the expulsion of the whey.
- The continuous stirring which accompanies the scalding stage further helps to expel whey from the curd, as does the subsequent 'texturizing' step (see below).

Once the desired scalding temperature has been reached, the curd/whey mix may be held at this temperature for some 30 minutes or more until the titrable acidity in the whey, i.e. lactic acid and acidity of the solids-not-fat, has reached the level associated with the cheese in question – around 0.2% lactic acid for Cheddar cheese. This so-called 'pitching' stage involves keeping the curd in suspension for long enough for the essential changes to take place, but once the curd has become firm – a parameter usually assessed subjectively by the cheesemaker – the whey is then drained off to leave a mass of curd in the bottom of the vat (Figure 1.13). Alternatively, the entire contents of the vat may be transferred to a special draining table so that the cheese vat can be cleaned, sanitized and refilled with fresh milk. Either way, it is essential that the curd develops a

Figure 1.13 After the 'pitching' stage, the whey is drained out of the tank and the curd is piled along the sides. (Courtesy of Group Rhône-Poulenc, France.)

specific texture prior to pressing. This texturizing stage – often referred to as 'cheddaring' because of its importance in the manufacture of Cheddar cheese – ensures that certain specific characteristics develop in the curd.

In traditional processes, the curd is piled along the sides of the vat to allow the small pieces of curd to coalesce into a solid mass. In modern factories, the same processes take place on a conveyor system which allows the curd to travel slowly from the draining tables to the finishing area. Where large blocks are formed along the sides of the vat, manual breakage to allow further draining of the whey is essential (Figure 1.14), but the degree of disturbance of the curd depends on the type of cheese being produced, i.e. a 'crumbly' cheese like Cheshire will require more extensive manipulation at this stage than a smooth textured cheese. Where a conveyor system is employed, the shallow bed of curd never coalesces to the same extent; hence, although some mechanical stirring of the curd may take

Figure 1.14 After the curd has been piled along the sides of the vat, the individual pieces gradually coalesce into a solid mass, which is then cut manually and stacked in order to improve whey drainage. (Courtesy of Group Rhône-Poulenc, France.)

place to encourage whey drainage, no other disturbance of the curd is required.

Handling of the curd: soft cheeses

The process for manufacturing soft cheeses like Cottage cheese or Coulommier, as well as the externally mould-ripened varieties such as Brie and Camembert, has much in common with the basic procedure until the formation of the coagulum. Beyond this point the processes for hard-pressed and soft cheeses diverge quite sharply.

Cutting of the coagulum is quite common. With some cheeses, like Cottage cheese, the pieces of curd are even scalded to provide particles of a reasonably firm texture. However, no procedure equivalent to texturizing is encountered, and removal of the whey is usually by drainage under gravity rather than mechanical pressure. With Cream cheeses which are homogeneous in struc-

ture, removal of the whey may well involve hanging in a muslin bag or centrifugation, but in most cases the soft curd is simply transferred to a mould. The effect of overnight draining is to give a lightly structured cheese, and one that retains sufficient moisture to be designated as 'soft'.

Next morning, the cheeses will be firm enough to be removed from the mould manually or automatically (Figure 1.15), prior to sale as fresh cheese or dry salting and maturation according to tradition. This maturation stage may involve the growth of external moulds, e.g. in the production of Brie, or bacterial cultures as with the manufacture of Brick or Limburger. Whatever the precise

Figure 1.15 After draining overnight, the individual cheeses are automatically discharged from the moulds for transfer to the salting/maturation stage. (Courtesy of Stork Amsterdam International Ltd, UK.)

procedure, the high moisture content of the base cheese is vital for success.

The same lack of structure implies that the packaging of the two types of cheese has to be approached in very different ways, e.g. polythene tubs for soft cheeses as against shrink-wrapped blocks for hard varieties. Distribution and storage/shelf-life considerations are also peculiar to the type of cheese in question.

The finishing stages: hard-pressed cheeses

Depending on the nature of the curd and its method of handling, the mass of pieces gradually acquire a recognizable texture. At this point, the curd is ready for salting and milling. With some varieties like Emmental, the salt content of the finished cheese is achieved by immersion in a brine bath, but most pressed cheeses are dry salted either before or after milling of the textured curd. The functions of salting, as discussed by Guinee and Fox (1987), are:

- The salt (sodium chloride) acts as a preservative, especially in the brined cheeses of Mediterranean and Middle Eastern origin.
- The flavour of the cheese is enhanced.
- The moisture content of the curd is reduced as the osmotic pressure exerted by the salt or brine draws out the water.
- The salt tends to inhibit the growth and/or metabolic activity of the starter bacteria which, ultimately, leads to lysis of the cells and the release of the enzymes contained therein. These enzymes, along with those from the coagulant, play a major role in the development of the flavour characteristic of the variety in question.
- Bacteria of non-starter origin are inhibited as well, hence the development of off-flavours and textural defects is avoided.

Although consumer preference can influence the level of salt added to a cheese, the correct development of a particular variety usually depends upon a precise level of salt being present. In Cheddar cheese, for example, the correct level

of salt in the finished cheese should be 1.7–1.8%, and batches with contents outside this range tend to be down-graded due to the risk of poor flavour development as maturation proceeds. The fact that salt content can be so critical means that even distribution throughout the mass of curd is vital, and one advantage of milling after salting is effective mixing. Milling, as the name suggests, involves cutting the now compressed curd into small pieces of a size optimum for the variety in question, prior to filling into moulds. The dimensions of the moulds will reflect the expected size and shape of the finished cheese, while the applied pressure is dependent on the nature of the cheese. Thus, cheeses with a low moisture content require a high pressure to expel the whey and consolidate the curd.

The duration of the pressing stage can also be varied to achieve a defined end-point, but usually

Figure 1.16 Blocks of Cheddar cheese being conveyed from the press to the packaging room, where each block will be individually wrapped and boxed prior to transfer to the maturation store.

Figure 1.17 Typical storage facility for the maturation of traditional Cheddar cheese. (Courtesy of the National Dairy Council, UK.)

a few hours for a semi-hard cheese to overnight for a hard cheese is sufficient to form a block of the desired consistency. After removal from the moulds, the cheese will be handled according to local custom but the process may well be fully automated in large-scale operations (Figure 1.16).

Packaging in barrier film or coating in wax often follows directly after pressing, with the objectives of:

- maintaining the shape of the cheese (further support from plastic or wooden boxes is often required for large cheeses of 20 kg or more in weight); and
- preventing the evaporation of moisture and the ingress of oxygen, so providing the correct conditions for maturation to take place at a controlled rate.

Given this degree of protection, the main demand upon the cheese store is control of temperature, and 5–10 °C is popular for many hard cheeses like Cheddar (Figure 1.17). For cheeses with 'eyes', like Edam or some of the Swiss varieties, a temperature of up to 20 °C may be employed to ensure that the bacteria responsible for gas production remain fully active.

Maturation

The nature and duration of the maturation process vary with the variety of cheese. Fresh, unripened cheeses may be sold a few days after manufacture but extra hard cheeses like Parmesan may be held for two years or more for the full flavour to develop. In other cases, like Brie and

Camembert, the growth of mould over the surface is essential for the biochemical processes of maturation to take place, and the same is true of the internal mould growth that ensures the unique character of Stilton or Roquefort.

Nevertheless, most varieties of cheese undergo a period of maturation prior to consumption. During this stage, the essential transformations usually involve:

- changes in flavour resulting from the activity of enzymes of coagulant or starter origin (Davies and Law, 1984);
- modifications to the structure of the cheese, often brought about by proteolytic enzymes originating from the starter culture or, in the case of mould-ripened cheeses, from the spreading mycelium of the fungus; and
- biochemical interactions associated with the extensive battery of compounds derived from the milk or the activity of various micro-organisms on these natural constituents.

The mature cheese is an extremely complex material, and judging the exact point at which a cheese has reached its optimum character is no easy matter. In general, the progress of maturation has to be judged subjectively by a skilled grader. For Cheddar cheese, for example, the process will entail:

- selection of a typical block or round of cheese at three months from the date of production;
- removing a core (Figure 1.18), some 6 in long and 1 in in diameter, and examining the sample for colour, aroma, texture/body and flavour;
- awarding a score for each attribute against a pre-determined maximum, weighted for relative importance (e.g. out of 40 for body and texture and 45 for flavour and aroma);
- totalling the scores so that the cheese can be placed into a specific Grade, e.g. 'Fine Grade' for cheeses scoring over 90 points, 'First Grade' over 80 or whatever point is chosen on

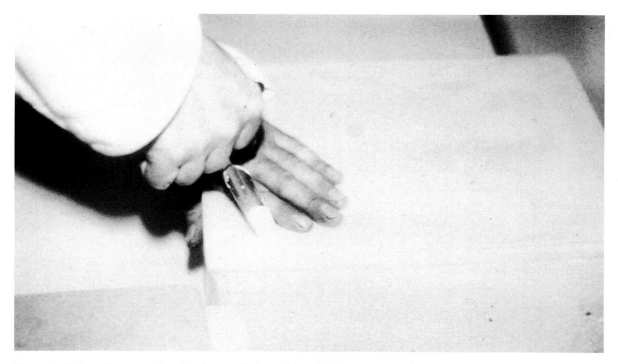

Figure 1.18 In order to grade a hard-pressed cheese like Cheddar, an expert will withdraw a cylindrical core with a cheese iron and then assess the quality before carefully replacing the extracted material into the hole.

the scale; 'Second Grade' is the lowest for general retail sale, and cheese that does not mature even into this lowest category may well be used for processing.

Although the maturation process is a natural biological and chemical one, it is usually assumed that cheeses made from the same vat will be ripening at roughly the same rate and in the same manner. Consequently, if one cheese is graded as 'First Grade', the remaining 40 or so cheeses from the same batch or vat will be given the same classification. For Mild Cheddar cheese, which may be sold at three months, this initial grading serves to determine the quality of the retail cheese, For mature cheese, which will be allowed up to 12 months or more to develop a full flavour, a second grading as the product nears maturity will be required to ensure that maturation has proceeded as expected.

This type of approach is applied in various ways to most cheeses. Although the timing and exact procedure will change in relation to the variety under test, a grading system is essential to ensure that:

- the quality standards of a given variety are maintained;
- the consumer is provided with a cheese that broadly matches the description; and
- the manufacturers of 'Fine Grade' cheese, by separating good quality cheese from the poor, are able to employ the classification to generate a higher income.

Indeed, with highly-prized and chemically complex cheeses like Stilton, each individual cheese may be graded, so that a top quality product will truly merit the description.

The above description of the manufacture and maturation of cheese is intended to be little more than an overview of the stages of production, for with some 900 named varieties on the world markets, and many more local derivatives of these recognized types, there is no such thing as a 'standard' process for the manufacture of cheese. When considerations of scale are added to the scenario, then clearly the priorities of a farm-house producer using the milk from a few cows *vis-à-vis* a major manufacturer handling millions of litres of milk will further complicate the issue. Nevertheless, the basic stages of manufacture remain immutable, and provide a basic framework against which individual varieties of cheese can be examined.

Fermented milks

Fermented milks are produced by the action of selected bacterial cultures on milk but, unlike the situation with cheeses, the liquid/whey phase is retained within the coagulated protein. Consequently, they are high-moisture products – usually in excess of 80%, with shelf-lives that reflect the abundance of water available to support the growth of contaminant yeasts or moulds. The low pH of most cultured milks (often in the region of pH 4.0) means that the conditions are too acidic for most spoilage bacteria to grow, and the same applies to potential pathogens. Consequently, fermented milks enjoy an excellent reputation for consumer safety and food-borne diseases like salmonellosis or listeriosis should never be linked with cultured products of this type.

The broad categories of fermented milks tend to be linked, as mentioned earlier, with geographical region and type of starter culture. The IDF (1988) grouped the products as:

- Mesophilic milks – fermented with lactococci and associated mainly with the peoples of Northern Europe.
- Thermophilic milks – produced by species of streptococci and lactobacilli that have optimum temperatures for growth and metabolism around 40 °C.

The principal difference between these groups of products, at least as perceived by the consumer, concerns the flavour imparted by the cultures. The basic procedures of manufacture have much in common.

Manufacture of fermented milks

The majority of milks on the commercial market are derived from cow's milk, but sheep's or goat's milk is used by manufacturers either to meet a special demand (e.g. by consumers who are allergic to bovine proteins) or if cow's milk is in short supply. However, whatever the exact raw material, the levels of milk solids, especially the non-fat components, are usually too low to give an end-product of acceptable quality in terms of texture and consistency. For fermented drinks like buttermilk, kefir or kumiss, the milk may well be used in its native state, but for 'spoon-able' products the first stage of manufacture will involve an increase in the total solids of the milk. The alternative techniques for achieving this aim are:

- Removal of some of the water by evaporation; traditionally, heating in a vessel over an open fire was the standard practice, but modern plants tend to use concentration under vacuum as the normal procedure. Removal of some of the water by membrane separation offers another alternative.
- Addition of non-fat milk solids, usually skim-milk powder (Figure 1.19) as it is stable at ambient temperature and hence can be transported and stored without costly facilities.

Figure 1.19 Although skim-milk powder is usually transported in multi-ply paper sacks lined with polythene, it must still be stored under clean, dry conditions if it is not to absorb moisture or undesirable taints.

Either way, the intention is to bring the solids-not-fat level (8.5–9.0% in normal milk) to:

- 16–18% for a natural set product, i.e. one with a defined gel structure;
- 13–14% for a stirred product with a noticeable viscosity and mouth-feel, but where the textural properties can be adjusted by the use of stabilizers, such as modified starch or various plant gums (e.g. guar gum or locust bean gum); or
- 11–12% for fermented products that may be used for cooking, or drinking yoghurts which need a more defined 'body' than can be attained from milk alone.

The adjustment of the fat content prior to fortification is at the discretion of the manufac-turer. While some markets are dominated by low fat products, levels of fat up to 10% can be found in some yoghurts and concentrated yoghurts like labneh; cultured creams or dips have fat contents more in keeping with market cream. This variability in fat content is influenced principally by consumer demand, in that whilst the fat may provide a perception of 'luxury' and improved mouth-feel, its presence is of marginal importance during manufacture; a degree of physico-chemical bonding between fat and protein may enhance the viscosity of some products, but in most cases the effect is unlikely to be dramatic. Obviously, high fat milks will have to be homogenized prior to fermentation to prevent the fat rising to the surface of a set retail product – an additional processing step that involves reducing the size of the fat globules by forcing the milk through a small orifice under high pressure. In the main, however, the properties of the end-product are dependent on the level and nature of the protein present.

The caseins are central to the formation of the coagulum and, in order to ensure their maximum activity in this respect, the fortified milk is next heated to 80–85 °C for 30 minutes or 90–95 °C for 5–10 minutes. The choice of time and temperature is mainly a reflection of the type of plant available, for if the product is being manufactured in one large, all-purpose vat (Figure 1.20), then the lower temperature range reduces the risk

Figure 1.20 A water-jacketed process vessel of the type that is often employed to manufacture fermented milks on a small scale (up to 1000 litres/batch). A built-in agitator is used for mixing; temperature control during heating, incubation and cooling is achieved by passing water at the correct temperature through the jacket.

of milk burning onto the sides of the vessel. However, if a plate heat exchanger is being employed (Figure 1.9), the high temperature/short holding time system is more convenient. The essential factor with either system is the extended holding time at an elevated temperature, because this treatment causes the whey proteins to denature and become, at least in part, adsorbed onto the casein. The end result is that, when the caseins coagulate during manufacture, a soft curd is formed that provides the basic qualities that will be sought in the retail item. In addition, the heating process will:

- eliminate any pathogens present in the raw material;
- drive out the oxygen from the milk, so making it a suitable environment for the growth of the starter bacteria (many lactic acid bacteria are micro-aerophilic, and grow and metabolize best under conditions of reduced oxygen tension).

Following the heating stage, the milk is cooled to the correct temperature for the fermentation (e.g. 22–25 °C for a mesophilic milk or 37–42 °C for a thermophilic milk) and inoculated with a culture appropriate for the end product. Beyond this point, two alternative routes are available (Figure 1.21). First, the inoculated milk may be transferred to retail cartons prior to incubation (set/gel products), or second, the milk may be incubated in bulk and then filled into cartons at a later stage (stirred products). Either way, the milk is held at the relevant temperature so that (a) the bacteria generate lactic acid and/or flavour compounds, and (b) the milk coagulates into the gel structure that is an essential characteristic of most products. The time taken to reach the desired end-point of around 1.0–1.4% lactic acid (pH 4.0–4.4) is principally determined by the temperature and/or choice of starter culture, so that whilst a mesophilic milk may be incubated overnight, a thermophilic culture may complete the fermentation in 3–4 hours.

At the end of this time, the milk will have coagulated into a homogeneous gel (Figure 1.22), the strength of which is determined in the main

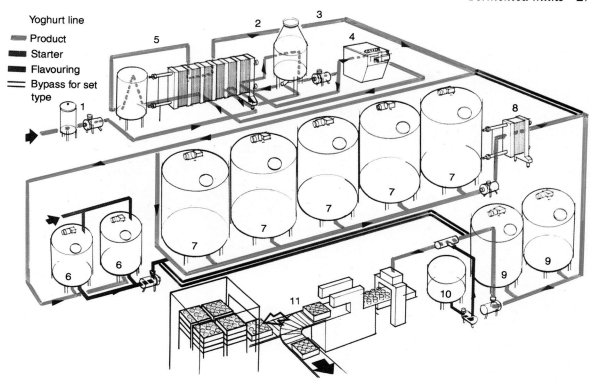

Figure 1.21 A typical line for production of set and/or stirred yoghurt. The fortified milk enters via the balance tank (1), and then passes through the plate heat exchanger (2) and holding tube (5), an optional vacuum chamber (3) to remove entrapped air and a homogenizer (4). It is then pumped either into the incubation tanks (7) or direct to the filling station (11); in either case, it will be inoculated with starter culture from the bulk tanks (6). The filled cartons are incubated to produce natural set yoghurt, while the yoghurt in the incubation tanks passes, after incubation, through the plate cooler (8) into the holding tanks (9) prior to the addition of fruit from a tank (10) and filling. All the retail cartons will be stored at 2–4 °C during distribution and sale. (Courtesy of Elsevier Science Publishers Ltd, UK.)

by the total solids of the original mix. In general, the gel can be visualized as a network of strands of protein. On stirring, this gel will break down into a series of small 'lumps' suspended in the fluid phase (Figure 1.22(a)). This mix has a sufficiently defined viscosity that the product would be acceptable to many consumers, but as the degree of agitation is increased, so the 'lumps' become smaller (Figure 1.22(b) and (c)) and the product more fluid. It is for this reason that stirred yoghurts require the presence of stabilizers like starch or a plant gum to retain their consumer appeal, for it is the viscosity of the fluid phase that is the critical influence.

A characteristic flavour will have developed also, with the important components deriving from the starter culture. As a consequence, mesophilic products tend to have a dominant 'buttery' note, derived from the activities of *Lact. lactis* biovar *diacetylactis* and *Leuconostoc* spp. leading to a build-up of diacetyl. Thermophilic milks often possess a flavour reminiscent of the best-known example, yoghurt; acetaldehyde is one of the main flavour components of the latter group.

After refrigeration, the milks may be consumed as 'natural' products – set or stirred – or fruits and flavours may be added (usually stirred pro-

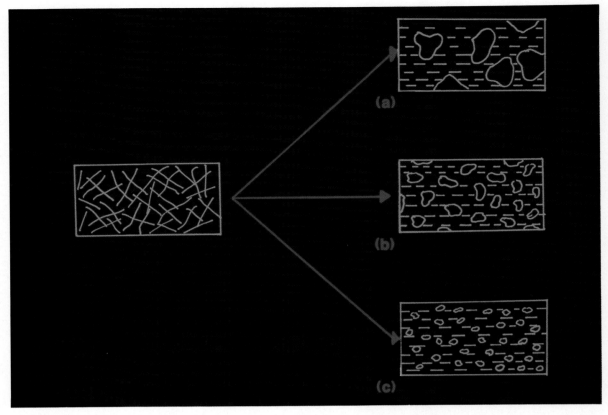

Figure 1.22 The coagulum of set yoghurt can be envisaged as a network of protein strands. When the gel is stirred, it breaks down into small 'lumps' whose size reduces ((a) → (b) → (c)) as the force applied to the system increases. (After Robinson (1988).)

ducts). Either way, the products will contain large numbers of bacteria of starter origin. These bacteria, which in well-made products will number several million per gram of product, are regarded as an essential property of a cultured milk and in many countries yoghurt can only be sold with the 'live' culture present. The reason for this insistence is:

- the traditional concept of a fermented milk;
- the evidence that viable bacteria in thermophilic products may offer prophylactic and/or therapeutic benefits to the consumer (see later);
- the flavour is more attractive than when the product is pasteurized after fermentation in order to eliminate the starter culture.

Consequently, there is every incentive for manufacturers to retain the fresh or 'live' image of fermented milks, and for consumers to demand compliance with this usual practice.

Although fermented milks should not suffer bacterial spoilage or permit the survival of pathogens, other than inactive spores, loss of quality can occur unless the products are held under refrigeration. The development of an excessively acid taste is one common problem – *Lac. bulgaricus*, for example, will continue to metabolize the residual lactose in a fermented milk until the level of lactic acid is almost double that anticipated, e.g. 2.4% as against 1.2%. Even at chill temperatures, this development can be a problem over a 2–3 week shelf-life, but above 10 °C the product may well become inedible in a matter of days.

Storage temperatures above 5 °C also encourage the growth of any yeasts or moulds that have entered the system, for unlike many bacteria, fungi are broadly tolerant of acid conditions. While moulds may do little more than produce an unwelcome colony on the surface of a set yoghurt, yeasts can cause much more damage. A typical spoilage yeast will not only generate off-flavours in a product, but also appreciable quantities of carbon dioxide. The build-up of this gas in heat-sealed cartons can prove disastrous in terms of leaking or 'blown' containers, and any manufacturer will take the threat very seriously indeed. Preservatives, such as sorbic acid, are permitted in some products; sorbic acid will control yeast growth but it has no effect on the viability of the desired bacterial flora. However, the use of such additives is frowned upon by many producers, leaving high standards of hygiene as the best protection against contamination.

2 Extra-hard cheeses

There are a number of cheeses produced, mainly in Mediterranean Europe, which can be consumed after a few months as a fresh cheese, but are more usually matured for two years or more to become hard and granular in texture. The flavour may be strong and harsh or delicate and faintly aromatic and, in the mature state, such cheeses are ideal for low level additions to other foods or for grating as a 'topping'.

Great care goes into the selection of the raw materials (usually sheep's or cow's milk) for manufacture and into ensuring that the maturation process follows the correct pathway, for the traditional process of manufacture is hundreds of years old. Italy is generally regarded as the centre of origin, and certainly the armies of Imperial Rome were familiar with these cheeses; it may be that their resistance to spoilage, associated with the low moisture contents of the group, made them ideal for transport by the legions.

Although the name Parmesan has become synonymous with good quality extra-hard cheese – at least outside Italy – a more accurate collective term would be Grana-type cheeses, i.e. cheeses with a hard, granular texture. According to Guinee and Fox (1987), the principal groupings for the extra-hard Italian cheeses are:

- Asiago.
- Parmesan/Grana types, including Parmigiano Reggiano, Bagozzo, Emiliano, Grana Lombardo, Grana Padano, Lodigiano, Modena and Monte.
- Romano types, including Romano, Sardo, Fiori Sardo and Pepato.

Many of the varieties have much in common and are recognized principally by the region of manufacture, e.g. Grana Lombardo from the province of Lombardy and Parmigiano Reggiano from Reggio in the province of Emilia, and many subvarieties have emerged in the same way. The use of sheep's milk rather than cow's milk provides a major contrast, and the distinctive Pecorino Romano from Sardinia owes much of its character to the ovine raw material. Even the introduction of modern technology has been carefully monitored to ensure the survival of essential traits, and a comparison of the traditional and modern systems for manufacturing Grana-type cheeses tends to confirm this view.

Asiago

Asiago is the name of a town in the Italian province of Vicenza and the original cheese, Pecorino di Asiago, was locally produced. Nowadays, although the name Asiago is recognized as a 'name of origin', a number of varieties have emerged, all based upon cow's milk.

A typical example, Asiago d'Allevo, is shown in Figure 2.1, and a high-fat variant is shown in Figure 2.2. However, while Asiago d'Allevo may be consumed fresh up to around six months old, it is more usually stored for longer periods for grating; Asiago Pressato is normally sold for table use.

The milk for Asiago d'Allevo is partly skimmed. At 33–35 °C, rennet paste – with or without additional lipase (Davis, 1976) – is mixed into the raw milk. This rennet paste, which is widely used in Mediterranean countries, is a somewhat crude form of the animal coagulants described in Chapter 1, and the unrefined mixture of enzymes tends to produce a cheese with a stronger, harsher flavour than could be achieved with more refined enzyme preparations. The employment of

Figure 2.1 Asiago d'Allevo, a medium-fat hard cheese from (mainly) the provinces of Vicenza and Trento. (Courtesy of the Ministero dell'Agricoltura e delle Foreste, Rome.)

Figure 2.2 Asiago Pressato, a high-fat hard cheese from (mainly) the provinces of Vicenza and Trento. (Courtesy of the Ministero dell'Agricoltura e delle Foreste, Rome.)

additional lipase to further break down some of the milk fat enhances this effect during maturation. Acidification of the milk with a locally prepared whey culture (described later) encourages coagulation, and the gel is cut into small pieces of some 0.5 cm maximum in dimension. The temperature in the vat is then raised to 45–47 °C and held for around 15 minutes. After most of the whey has been drained off, the curds are transferred to circular hoops 30–36 cm in diameter, in quantities that give finished cheeses of 8–12 kg. After the warm curd has compacted in the hoops, the cheeses are pressed overnight before being transferred to a brine bath where they will remain for several days.

Maturation at ambient temperature follows. If not coated with wax, the cheeses will be turned at regular intervals and rubbed with olive oil to build up a smooth, flexible rind. At around six months, the cheese will have developed a compact texture with a scattering of small to medium-sized 'eyes' and will have a mild flavour. However, as maturation proceeds, the texture becomes hard and granular, and the flavour strenghens as the result of continuing enzyme activity; at 12 months, it will be suitable only for grating.

Typical compositions of the two varieties of Asiago are shown in Table 2.1.

A type of Asiago is now manufactured in the USA. According to the USDA (1974), three types can be recognized:

- Asiago (fresh) – matured for at least 3–4 months, and having a moisture content of 40–45% and a fat-in-dry-matter (FDM) of not less than 50%.
- Asiago (medium) – matured for at least 6 months, and having a moisture content of <35% and an FDM of >45%.
- Asiago (old) – matured for at least one year, so that the moisture level is not above 32% and the FDM >42%.

Table 2.1 Typical compositions (%) of the two varieties of Asiago

	Asiago d'Allevo	Asiago Pressato
Protein	28	24
Fat	30	30
Fat-in-dry-matter	44	48
Moisture	33	40
Salt	2.2–2.7	1.5–1.8

Note: the fat content of Asiago d'Allevo depends on the degree of skimming of the original milk.

Caciocavallo

The name of this cheese is alleged to be derived from the fact that, during the ripening stage, cheeses are tied in pairs over long wooden poles, i.e. the cheese (*cacio*) is placed astride (*a cavallo*) a support. However, as the origins of this cheese go back several hundred years, the validity of this tale is hard to verify, as is the historical use of the cheese – was it produced for eating fresh or for prolonged storage and use as a grating cheese? Nowadays, it is accepted that Caciocavallo can be sold young or old as desired, with maturation times ranging from 2–12 months or even longer.

The fresh cheese in Figure 2.3 shows the typical pale, compact interior of the product along with the distinctive golden-yellow rind. Equally traditional is the shape, which is round/oblong with a short neck and ball-shaped top. The weight is around 2 kg.

Although mixtures of sheep's and cow's milk were originally used for production in southern Italy and latterly in Sicily, full cream cow's milk is now the designated raw material. The method of manufacture, and indeed the fresh cheese itself,

has much in common with Provolone (see page 82–3), except that smoking of Caciocavallo is not so common. However, kneading the curd in hot water is essential to give the smooth, slightly pliable curd, and brining for 3–4 days at 10 °C is a standard practice. After their surfaces have been dried, the cheeses are hung in pairs in time-honoured fashion to mature, and cleaned with olive oil from time to time to ensure the development of a smooth, unblemished rind. The minimum ripening time is usually 3–4 months (or perhaps 6 months for a slightly stronger flavour). Extended periods of 12 months or more are only employed for cheeses destined for grating. The flavour of the young cheese is mild and delicate, but age brings with it a rich, fragrant flavour that is much sought after for cooking.

The composition of a cheese at around 6 months might be:

Protein	30%
Fat	30%
Fat-in-dry-matter	44% (minimum)
Moisture	30%

Figure 2.3 Caciocavallo, a medium-hard cheese produced in central southern Italy from cow's milk. (Courtesy of the Ministero dell'Agricoltura e delle Foreste, Rome.)

Montasio

As with a number of hard cheeses, the characteristics of the product change with time of maturation and Montasio can be consumed as a fresh cheese within 2–3 months of production, or employed as a condiment/grating cheese after 14–18 months.

Originally, the name tended to be applied to any cheese produced in the region outlined in Figure 2.4, and cow's, goat's or sheep's milk was employed as the starting material. More recently, the name has achieved official status and a 'Consorzio per la Tutela del Formaggio Montasio' has been established to ensure that all producers comply with certain regulations; only cheeses marked with the official 'logo' (Figure 2.5) should be described as 'Montasio'.

Consequently, good quality raw cow's milk has become the starting point for the process, and it

is anticipated that the natural microflora from the milk and/or the factory will provide sufficient acidity for the calf rennet to achieve coagulation. A slightly elevated ripening temperature of 33 °C is employed to encourage gel formation within 40–60 minutes and at this point the curd is cut into pea-sized granules. Slowly raising the temperature of the stirred curd/whey mixture to 46 °C produces firm pieces of curd that rapidly settle from the whey on cessation of the stirring. Once some of the whey has been decanted from the vat, portions of the curd are collected into cloths and placed in moulds of some 30–40 cm in diameter and 8–10 cm in height. Pressing over the next day, with frequent turning, forms a cheese that is sufficiently firm to receive the first application of dry salt.

Further salting takes place in an ambient store at 10–14 °C. By the end of the first month, the salt content of the cheese should be around 1.5%.

Figure 2.4 The names of many cheeses enjoy protection against application to cheeses other than the traditional variety. In this case, Montasio has to be produced within the area shown. (Courtesy of Consorzio per la Tutela del Formaggio Montasio, Italy.)

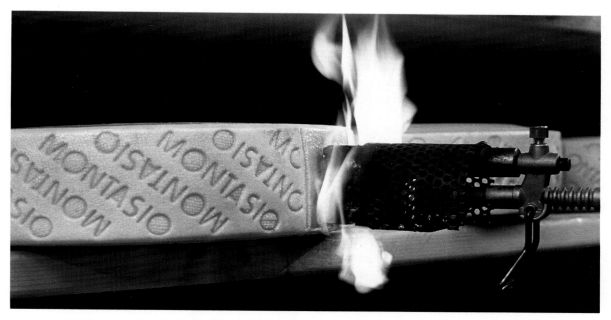

Figure 2.5 A Montasio cheese produced according to the appropriate regulations is eligible to have the distinctive 'trade-mark' burnt on to the side. (Courtesy of Consorzio per la Tutela del Formaggio Montasio, Italy.)

Further maturation continues at ambient temperature and a humidity of 80–85%, and the consequent drying produces a cheese of around 36% moisture after two months. At this point, the cheese will have a mild flavour and a pale colour (Figure 2.6) and can be consumed fresh, but much cheese is allowed to mature for a longer period. With the passage of time, the colour of the interior darkens, the flavour becomes more piquant and the small eyes that are another characteristic of mature Montasio become more evident.

The typical composition of the cheese at around 12 months is:

Protein	28–30%
Fat	32–34%
Fat-in-dry-matter	45–47%
Moisture	27–29%
Salt	1.95%

From this stage onwards, grating is the usual form of consumption.

Figure 2.6 As Montasio matures, so the colour darkens and the texture hardens, and the fresh table cheese (bottom) changes into the grating variant (top). (Courtesy of Consorzio per la Tutela del Formaggio Montasio, Italy.)

Parmesan (Grana) cheeses

Grana Padano (Figure 2.7) and Parmigiano Reggiano (Figure 2.8) are probably the most important varieties in this group, but many towns have their names associated with cheeses of local production. The differences between the cheeses are often difficult for the uninitiated to appreciate, but subtle contrasts with regard to the type of feed available to the cows, for example, can be discerned by the connoisseur. In some cases, the retention of special characteristics is reinforced by decree: for example, Parmigiano Reggiano can only be manufactured between 15 April and 11 November. This limitation means that the cows will be feeding, in the main, on clover or lucerne, and hence the quality of milk will be excellent. Nevertheless, the basic procedures of production for all grana cheeses have much in common, and they are essentially a blend of techniques – traditional and modern.

In the original system, raw milk drawn on the evening before production was poured into large

Figure 2.8 Parmigiano Reggiano, one of the classic medium-fat hard cheeses of Italy. (Courtesy of the Ministero dell'Agricoltura e delle Foreste, Rome.)

Figure 2.7 Grana Padano, a hard, medium-fat, slow-ripening cheese produced in many provinces of Italy. (Courtesy of the Ministero dell'Agricoltura e delle Foreste, Rome.)

shallow vats and left overnight. Part of the separated cream was removed next morning, and fresh milk was added to give a total volume of some 500 litres of partly skimmed milk; 500 litres being the volume required to produce one cheese of 23–38 kg (Bottazzi, 1993). The milk was then poured into a copper vat and the natural microflora was allowed to develop the required acidity. A fairly crude calf rennet was used to obtain the coagulum, which was cut into small pieces before being exposed, with continual stirring, to a scalding temperature of 53–55 °C (Figure 2.9). The curd, when it was sufficiently firm, was allowed to separate either under gravity or with the assistance of a light weight. The curd was then scooped into a cloth drawn across the

Figure 2.9 Traditionally, the initial stages for making Parmigiano Reggiano were carried out in copper kettles formed into the shape of an inverted church bell. (Courtesy of Leo Bertozzi, Consorzio del Formaggio 'Parmigiano Reggiano', Italy.)

was taken to avoid cracks developing in or mould growth over the developing rind. Olive oil would have been the most commonly employed material to rub the coats, but in some regions mixtures of charcoal and grape seed oil were used, especially during the later stages of maturation, to give the cheese a black, shiny rind (see Pecorino Romano).

Even today, the manufacture of some types of Grana, such as Parmigiano Reggiano, differs little from the original process, but the producers of Grana Padano in particular have sought to update their procedures. The partial skimming of the milk, for example, often takes place in bulk rather than as a two-step process, and the use of starter cultures is common. Some factories use a standard yoghurt starter of *Str. thermophilus* and *Lac. bulgaricus*, but so-called 'whey cultures' are still in common use. As the names suggests, they are cultures grown by inoculating a volume of fresh whey with a culture from a previous day, and the resultant microflora is by no means well defined. It is usually a mixture of thermophilic lactobacilli, such as *Lac. helveticus*, *Lac. bulgaricus* and *Lac. lactis*, and its special features are:

- long association with the dairy has rendered the strains of bacteria resistant to any resident phages;
- the strains rapidly produce lactic acid in milk;
- the cultures are cheap and easy to prepare; and
- the complex nature of the culture may lead to cheeses with interesting flavour profiles.

Beyond the use of starter cultures, the main concession to 'technology' is an increase in vat size to 1000 litres, i.e. sufficient milk to manufacture two cheeses per vat. Some mechanization/standardization of the maturation systems has been achieved, e.g. the turning and cleaning of the cheeses, and modern cheese stores are held at 16–18 °C and a humidity of 85%. The movement of air throughout the store is also carefully controlled, but otherwise the production of Grana cheeses runs along traditional lines dictated by the characteristics desired in the end-product. The length of ripening varies with the cheese, so that while Grana Padano may be ready for sale in 13–15 months, the maturation span

bottom of the vat and, by pulling together the corners of the cloth, the mass of curd could then be lifted out. After hanging so that the free whey could drain out, the curd – still in the cloth – was transferred to a mould of dimensions appropriate to the variety, and a wooden board was placed on top. Light pressure along with frequent turning led to further whey being expressed, until finally the cloth could be removed and the moulded cheese exposed to pressure overnight. On removal from the moulds, the cheeses were then salted in brine baths for around one month. The finished cheeses were stored at ambient temperature for up to two years, and much care for Parmigiano Reggiano is 18–24 months (Figure 2.10). The typical compositions of the cheeses are given in Table 2.2.

Being large cheeses of some 35–45 cm in diameter and 18–25 cm high, cutting the mature cheeses for retail sale requires considerable skill. Special short-bladed, lance-shaped knives are used, and the first step involves tracing a line along the top and sides of a cheese. This line should divide the cheese into exact halves, and the knife should not penetrate beyond 1–2 cm. On fracturing the cheese, the natural, granular structure is retained (Figure 2.11). This approach has to be employed for all subsequent subdivisions so that there is an exact proportion of rind to cheese, and knife marks do not scar the surface

Table 2.2 Typical compositions (%) of Parmesan (Grana) cheeses

	Grana Padano	Parmigiano Reggiano
Protein	35	33
Fat	28	28
Fat-in-dry-matter	32	41
Moisture	30	30

of the cheese. Consequently, the consumer will be presented with portions (Figure 2.12) that have thick golden rind of 5–6 mm, a pale straw colour, and granular, flaky surface that may reveal cracks but few holes.

Figure 2.10 The classic method of checking the maturity of Parmigiano Reggiano is to tap the whole cheese with a small hammer and listen for the correct note. (Courtesy of Leo Bertozzi, Consorzio del Formaggio 'Parmigiano Reggiano', Italy.)

Figure 2.11 The method of scoring the mature cheese and breaking it into units ensures that all pieces have a rough, fractured surface. (Courtesy of Leo Bertozzi, Consorzio del Formaggio 'Parmigiano Reggiano', Italy.)

Figure 2.12 A mature Grano Padano cheese undergoing division into retail portions. (Courtesy of Consorzio per la Tutela del Formaggio Grana Padano, Italy.)

Ragusano

This cheese is manufactured rather in the manner of Caciocavallo and, although the name is protected under Italian law, it is not widely known outside its region of production – Sicily.

Full cream cow's milk is employed for manufacture, and the finished cheeses vary widely in weight from 6 to 14 kg depending on the wishes of the producer and the demands of the market; cheeses sold locally tend to be smaller than those transported to more distant markets.

Young cheeses, which may be consumed as table cheese, have an extremely mild flavour and a thin, smooth, golden-yellow rind, but the grating cheese (Figure 2.13) has a more pro-

Figure 2.13 Ragusano, a hard (usually grating) cheese produced in Sicily from cow's milk. (Courtesy of the Ministero dell'Agricoltura e delle Foreste, Rome.)

nounced flavour and a much darker rind. The texture is uniform with few, if any, holes visible and, with age, it becomes more compact and liable to fracture.

The typical composition of Ragusano is as follows:

Protein	27%
Fat	29%
Fat-in-dry-matter	44% (min.)
Moisture	31%

Reggianito

Argentina has long enjoyed close links with Europe, and hence it is perhaps not suprising that it is second only to the USA as a cheese producer in the entire American continent (Carlos Zalazar and Carlos Meinardi, personal communication). In 1991, for example, around six million tonnes of raw milk were produced, and some 47% went into the production of cheese. The end-products tend to fall into three categories: hard/grating cheeses like Reggianito, Sbrinz and Sardo (40 000 tonnes in 1990); semi-hard cheeses such as Holanda (Pategras), Colonia and Barra (96 500 tonnes in 1990); and soft cheeses (129 000 tonnes in 1990), of which Cremoso is the most famous.

The European origin of these varieties is self-evident in many cases, but differences in the composition and properties of the milk supply, together with subtle differences in the conditions of manufacture, conspire to produce cheeses that are distinctly Argentinian. The use of starter cultures of local origin tends also to reinforce the differences between the European and South American cheeses.

Reggianito is a good example, for not only does it contain more fat and moisture than comparable varieties from Italy, but also the use of a natural whey starter (often together with pancreatic lipase) provides a quite distinctive character to the product. It is normally a round cheese of some 7 kg, and the surface is coated with black paint as a protection against the adventitious growth of moulds (Figure 2.14). Although some cheese is consumed fresh, the majority is stored for around nine months until a strong, piquant flavour has developed along with a crumbly texture.

Figure 2.14 Reggianito, a very hard variety that is used both as a table cheese and for grating. (Courtesy of Dr Carlos Zalazar, Instituto de Lactologia Industrial, Argentina.)

Figure 2.15 Sardo, a very hard variety similar to Reggianito but smaller and matured for a shorter period. (Courtesy of Dr Carlos Zalazar, Instituto de Lactologia Industrial, Argentina.)

One unusual feature of the maturation stage is that the cheeses are 'graded' after six months, and only those of high quality are allowed to remain in store for the full period. Any cheeses with trivial defects are labelled as Sbrinz cheese and sold on the local market, whilst the remainder of the 'rejects' are grated, and the dried powder sold for cooking purposes in packets of 30 g or 160 g.

A cheese that closely resembles Reggianito in many respects is Sardo (Figure 2.15), which is made in the same manner but is formed into cheese of around 3.5 kg. It is ripened much more quickly than Reggianito (4–5 months in all) and so although hard in texture, it tends not to fracture in the classic manner of grating cheeses. The flavour is strong and piquant. Whilst the rind is often left natural in colour, coating with black paint is not uncommon.

Romano varieties

Most of the cheeses in this group are made from full fat sheep's milk; hence the prefix 'pecorino' (sheep) is common. Varieties made with cow's milk (Vacchino Romano) or goat's milk (Caprino Romano) are available but they do not really compete in terms of quality with the original. Although a mainland cheese in the first instance, Sardinia has become a major producer and exporter in recent years, and 'Sardo' has now become the accepted name for the island product. The raw material and location of factory are the principal determinants of the distinctive flavour, but adherence to the traditional techniques of manufacture is also a contributory factor.

The major period of manufacture depends on the availability of milk, and November to June is the important season for Pecorino Romano (Figure 2.16). Although raw milk was preferred by early cheesemakers, the procedure is now one in which the milk is collected, filtered, standardized to the desired fat content (around 2.0%) and then pasteurized. After cooling to 35–37 °C, the bulk milk is inoculated with either a whey starter, or a yoghurt starter of *Str. thermophilus* and *Lac. bulgaricus* at around the 1% level (or equivalent). Once microbial activity has been initiated, a crude lamb rennet paste is added which provides

Figure 2.16 Pecorino Romano, a classic hard cheese of Italy made from ewe's milk. (Courtesy of the Ministero dell'Agricoltura e delle Foreste, Rome.)

Figure 2.17 Pecorino Romano is used mainly as a grating cheese to provide a distinctive flavour in speciality dishes. (Courtesy of the Wisconsin Milk Marketing Board.)

not only the essential proteolytic enzymes but also lipases, which eventually will help to develop the strong, slightly rancid flavour desired in the finished cheese. The resultant coagulum is cut into pieces of a few millimetres in overall dimension, and then scalded at 45 °C for up to one hour.

As this temperature is optimum for *Lac. bulgaricus*, acid development continues, but the active involvement of the less thermo-tolerant streptococci will decline with time. In addition the curd particles take on a firm structure and, once stirring ceases, a mass of curd builds up on the bottom of the vat. Around 30 minutes later, the whey is run off to leave a coalescing mound of curd that is then broken manually into coarse lumps and placed in moulds. A period of natural drainage is followed by the application of light pressure for 1–2 hours, after which the cheeses are transferred to the salting room.

Although immersion in brine for 2–3 days is practised in some factories, regular dry salting at 15 °C over a period of 3–4 months is normal for all cheeses, including those that are brined first. The final maturation stage takes place in cool cellars or similar rooms at 8–10 °C. The minimum maturation time for Pecorino Romano is 8 months; during this time, the cheeses are regularly washed in brine and then coated with olive oil.

Carbon black may be employed in the coating to provide a distinctive appearance (Figure 2.17), but various hues from pale grey/white through to dark brown are equally acceptable. The interior of the cheese may be white through to pale straw in colour, but the texture should always be hard and granular. In size, the cheeses vary from 10 to 20 kg in weight, and some typical compositions are given in Table 2.3.

While Pecorino Romano finds use only as a condiment, the classic cheese from Sardinia, Fiore Sardo, can be eaten when young as a table cheese. Its mild flavour and soft texture – a reflection of the higher moisture content and lower scalding temperature – make it suitable for eating from one month onwards. However, 5–6 months of maturation give rise to a strong flavour and much

Table 2.3 Typical compositions (%) of Romano varieties

	Pecorino Romano	Fiore Sardo
Protein	27	25
Fat	29	29
Fat-in-dry-matter	36 (min.)	40 (min.)
Moisture	31	41

Figure 2.18 Fiore Sardo, a hard cheese produced from ewe's milk in Sardinia. (Courtesy of the Ministero dell'Agricoltura e delle Foreste, Rome.)

harder texture (Figure 2.18), so that from this point on it is seen essentially as a grating cheese.

Closely related to these varieties is Pecorino Siciliano which, as the name suggests, is produced

Figure 2.19 Pecorino Siciliano, a hard, usually grating, cheese made from ewe's milk in Sicily. (Courtesy of the Ministero dell'Agricoltura e delle Foreste, Rome.)

from sheep's milk on the island of Sicily. It is a cylindrical cheese up to 18 cm in height and 12 kg in weight – depending upon the producer – and the white, oiled surface often carries an imprint from the basket in which it was pressed (Figure 2.19). The texture is hard and compact with few visible fissures and, although sometimes eaten fresh, the strong flavour that develops after 4–5 months makes it more suitable for grating. The moisture content is usually in the region of 42%, and the minimum FDM has been set at 40%. A similar cheese can be found on the market with a special pepper added to enhance the taste, and such cheeses can be recognized in that they should have a definite 'cone' shape.

Sapsago

Sapsago cheese is one of the most unusual cheeses to originate in Switzerland and, according to the USDA (1974), it has a history that can be traced back over 500 years. It is a small, dry, cone-shaped cheese weighing between 0.5 and 1.0 kg, and its special feature is the sharp, pungent flavour originating from the incorporation of clover leaves.

It is made from slightly sour, skimmed milk, which is poured into a traditional round kettle and heated to near boiling. Cold buttermilk is added slowly to the hot, stirred milk, and any coagulated material that floats to the surface is skimmed off for later use. Sour whey is then added to bring about an acid coagulation of the casein fraction, so that the curd settles out when the stirring is stopped. The curds are then collected into a cloth or strainer, and spread out to cool and drain. Once at ambient temperature, the coagulated material that had been skimmed off earlier is mixed back into the curd, along with salt, and the mass is transferred to perforated wooden moulds. Each mould is then covered with a lid, and subjected to a heavy pressure overnight. Over the next 4–5 weeks, the moulds are held under a light pressure – still at an ambient temperature of 15–16 °C – and during this time, the initial maturation of the curd takes place.

By this time, the curd is in a comparatively dry

condition and the individual 'cheeses' may well be transported in sacks or barrels to another factory for final processing. This last stage involves grinding up the dry 'cheeses' along with 5% salt and 2.5% dried, powdered leaves of the aromatic clover, *Melilotus coerulea*. This mixture is stirred to give a smooth, homogeneous paste, and then packed into cone-shaped moulds (10 cm high, with sides tapering from a 75 cm base to a 50 cm apex) and lined with cloth. Once consolidated, the now-finished cheeses are tapped from the moulds. The clover provides the cheese with a pleasant, sage-green colour and attractive aroma, and it is much sought after as a condiment.

A typical analysis of Sapsago might be:

Protein	40–42%
Fat	5–9%
Moisture	38%
Salt	4–5%

Both the salt and the low moisture make for an extremely stable product.

Sbrinz

Sbrinz is a grating cheese that is currently being promoted outside Switzerland with some success (Figure 2.20). It is a very hard cheese with a grainy texture and sharp, nut-like flavour (USDA, 1974). It is a large, flat cheese, often being over 50 cm in diameter but only 10–15 cm thick, and it may be stored for up to three years before sale. Like many grating varieties, some of the cheeses may be sold much earlier as fresh cheese, and such products usually appear on the market under the name of Spalen cheese.

As it is often made in small factories, the method of manufacture and the end-products can show considerable variation but, even so, certain stages of the process tend to be followed in all plants. Raw, full-cream milk is the usual starting point and, after adjustment of the temperature to 25–26 °C, a sour whey starter culture is added. Standard rennet (25 ml/100 litres of milk) ensures that coagulation occurs within 40–45 minutes, a process that may well be aided by residual enzyme activity introduced with the whey starter.

Cutting into 2–3 mm cubes is achieved in a manner that varies with the shape and size of the vat, and stirring during the subsequent scalding stage can be mechanical or manual depending upon the premises.

Scalding over the next hour brings the final temperature to around 57 °C (Scott, 1986), at which point a volume of cold water is added to standardize the temperature at 55 °C and to dilute the whey being expelled from the curd pieces. Final firming of the curd takes place over the next 30 minutes, and the whey is drained off. The curd is then transferred to moulds of the appropriate dimension, and pressed for 2–3 days at a pressure that depends upon the plant available. After removal from the moulds, dry salting, brining or a mixture of both processes is pursued over the next 2–3 weeks, until the cheeses are dried and stored at ambient temperature for a further period. During this time, the cheeses are wiped at frequent intervals with a brine-soaked cloth, and also coated in linseed oil. Long-term

Figure 2.20 Although some Sbrinz is sold early for direct consumption, most is matured for 2–3 years and sold for culinary use as a grating cheese. (Courtesy of 'Cheeses from Switzerland'.)

storage for up to three years under cool conditions follows, during which time the cheeses may be turned occasionally and oiled.

A typical analysis for Sbrinz might be:

Protein	31%
Fat	34%
Fat-in-dry-matter	47% (min.)
Moisture	28%
Salt	4–5%

3 Hard-pressed cheeses

In terms of commercial production, the cheeses in this group are amongst the most important world-wide, and most dairy countries manufacture at least one cheese in this category. The basic character is determined by the moisture content, usually between 30% and 45%, and the fact that many of the cheeses are subject to considerable pressure during manufacture. The end result is that the final cheese has a firm, even texture with no more than a few mechanical fissures to disturb the appearance. When young, many extra-hard cheeses have rather similar properties and may be consumed as normal fresh cheese, for only at full maturity does their hard, granular nature become apparent.

The basic processes for making hard-pressed cheeses have much in common, and the important stages can be summarized as follows:

- Renneting and coagulation at moderate temperatures, e.g. 30 °C.
- Cutting the coagulum into small pieces, followed by scalding at a temperature of 39–40 °C.
- Removal of the whey.
- A period of in-system texturization of the curd (cheddaring) to allow further development of acid and desired changes in the structure of the casein.
- Milling of the dry curd at an acidity determined by the type of cheese.
- Subjecting the milled and salted curds to a high pressure for around 12–16 hours or longer.
- In-store maturation for periods of 3–12 months (or even longer) depending upon the degree of maturation sought for the finished cheese.

According to Lawrence and Gilles (1987), this type of cheese was originally developed in England, probably as a response to the cold climate. By piling the curds into 'heaps' whilst the whey was draining, the centre of the mass of curd would have been insulated against heat loss, and fermentation essential to the character of the cheese would have been able to continue. The advantage of ensuring the continued production of lactic acid is the reduction of the pH of the curd to below 5.4, at which point not only is activity of potential spoilage organisms reduced, but also the desired changes in the nature of the casein, involving a loss of calcium from the micelles, can occur. These essential alterations in the casein are common to the production of all hard and semi-hard cheeses, and the real difference between the types is the extent to which the changes are allowed to proceed, i.e. as the pH of the young cheese falls, so the calcium in the curd is reduced and the texture becomes increasingly 'crumbly'.

Although rather limited use is made of the chemical composition of a cheese in relation to its identity as a specific variety, certain legal requirements may have to be met. Cheddar cheese, for example, is allowed a maximum moisture content – calculated on the total weight of the cheese – of 39%, whilst the comparable figure for Cheshire cheese is 44%. Fat content is closely monitored also, and for most cheeses in this group the minimum percentage of FDM is 48%; on a total weight of cheese basis, this value gives a minimum figure of 29% for Cheddar cheese.

Whilst the fat is monitored for legal purposes, the level in the cheese milk or, more specifically, the fat:casein ratio is important to the cheese-maker. The usual value for a Cheddar-style cheese is between 0.67 and 0.72, and this figure is attained by standardizing the bulk milk prior to use (Figure 1.9). If the milk is not standardized, then problems in processing may well be encountered. In particular, high fat levels make it difficult to remove moisture, so giving rise to a soft curd that will never develop the textural properties expected of the final cheese; a cheese with a high protein content will tend to be dry and have a 'mealy' texture.

The starter cultures employed for this type of cheese tend to be standard ones of *Lactococcus lactis* subspp. *lactis* and *cremoris*, with the aroma-producing *Leuconostoc* sp. and *Lact. lactis* biovar *diacetylactis* being added on occasion to provide a more complex flavour. The form of the starter may include any one of alternatives described in Chapter 1, and the basic process of cheesemaking follows the guidelines indicated above. It is, therefore, subtle differences in the handling of the curd and choice of starter that ensure the desired end-product is attained.

Cantal

Cantal is a hard cheese, not unlike Cheddar in many respects, that is named after Cantal in the Massif Central area of France. Milk from the high pastures of the Auvergne is reported to provide the best cheeses, and they are tall (up to 50 cm) and cylindrical (diameter up to 40 cm) in shape. The sheer size of the mature cheeses – 30–45 kg in weight – makes them more suitable for factory than farm production; hence it is not surprising that the farmhouse cheese, Le Cantalet, is around 10 kg (Figure 3.1).

Although some farmhouse cheese may still be made with raw cow's milk and no starter culture, factory production involves standardizing the milk to around 3.0% fat and then pasteurizing at 71 °C for 16–20 seconds. On cooling to 30 °C, about 1.0% of an active mesophilic starter is added, along with standard rennet (25 ml/100 litres of milk) and often a low level of calcium chloride as well. This latter material is added to ensure that a firm coagulum is produced because,

Figure 3.1 Cantal, a hard-pressed cheese from the Auvergne, closely resembles Cheddar cheese. Le Cantalet is a 'baby' farmhouse version. (Courtesy of Texel, Epernon.)

unlike with many hard cheeses, there is no scalding stage to modify the curd as it develops. About 40–60 minutes later, the gel is firm enough for cutting into cubes of 1 cm³, and the mass of curds is stirred gently for 20–30 minutes at 30 °C to assist with the release of the whey. The whey is then drawn off to leave the settled curds on the bottom of the tank. As the curd begins to coalesce into a solid mass, it is broken and turned at regular intervals, and weights are often placed upon the matted curd to expel more of the trapped whey. After some 24 hours of this treatment, either in the vat or on a separate draining table, the curd has developed sufficient acidity for it to be milled and salted to around 2.0%.

In farmhouse systems, it is sometimes allowed to stand for a further period after milling, but more usually the salted curd is filled directly into perforated, open-ended, steel moulds lined with cheese cloth. Followers placed at each end of the mould ensure that the applied pressure (up to 40 kg/kg of cheese) is transmitted across the full diameter of the cheese, and the mould may be turned 3–5 times during the pressing stage. Finally, the cheeses are removed from the moulds and cloths, and placed in a store at 8–10 °C for maturation. Three to six months represents a typical in-store period, and occasional turning and wiping with a cloth soaked in brine are desirable to maintain the rind in a clean condition.

A reddish-yellow rind is typical of Cantal, and the interior of the cheese will be pale yellow. The texture is firm and pliable when young, but becomes slightly more 'breakable' with age; a few natural fissures may be evident, but no gas holes. The flavour has been described as piquant with a hint of bitterness (Scott, 1986) but the best cheese is alleged to possess an aroma typical of a summer pasture.

A typical analysis of the mature cheese might be:

Protein	20–25%
Fat	28–34%
Moisture	40–45%
Salt	2–3%

Cheddar cheese

Cheddar cheese is the classic, hard-pressed cheese. From its origins in Somerset, England, in the 16th century, it has become a cheese that is manufactured throughout the English-speaking world. It varies in colour from pale to deep yellow, and the flavour ranges from mild and creamy (Mild Cheddar) to strong and biting (Mature Cheddar). These differences reflect, more than anything, the age of the cheese, for whilst Mild Cheddar can be sold at around 3–4 months, an extended storage period of 12–24 months will give a 'mature' flavour of some intensity. The texture of the mild form is close and firm but, because of the chemical changes that take place during 'cheddaring', slices cut from a cheese are distinctly pliable. At full maturity, the cheese tends to become somewhat harder.

The decision to manufacture either a mature or a mild product is often a commercial one – e.g. Farmhouse Cheddar cheese (Figure 3.2) is likely to be a mature variant – but the decision can also be based upon the results of a preliminary grading at 2–3 months after manufacture; natural variations between batches mean that not every cheese will be suitable for extended maturation.

Pasteurized full-cream milk is used for Cheddar cheese. To ensure the correct properties for maturation, the fat:casein ratio must be standardized to 0.67–0.72 (Lawrence and Gilles, 1987). The starter culture is usually a mixture of *Lactococcus lactis* subsp. *lactis* and *Lact. lactis* subsp. *cremoris*. For a more complex flavour profile, low levels of *Lact. lactis* biovar *diacetylactis* and *Leuconostoc mesenteroides* subsp. *cremoris* may be included as well. After ripening the milk to an acidity of 0.20–0.22% (as lactic acid), renneting and, finally, cutting into cubes (0.3–0.5 cm³), the curds and whey are slowly heated to around 39 °C. The selection of this temperature for scalding is based upon the desire to:

Figure 3.2 English Farmhouse Cheddar cheese. Variants of this product are now made world-wide but the traditional cylindrical cheese has proved less popular than the 'block'. (Courtesy of The Farmhouse English Cheese Bureau, UK.)

- reduce the numbers of starter bacteria in the curd by killing some of the heat-sensitive strains of *Lact. lactis* subsp. *lactis* (if excessively high numbers are present in the final cheese, bitter off-flavours may develop during maturation); and
- cause the 'cubes' of curd to contract under the combined influences of heat and acidity, so expelling much of the entrapped whey.

Once the granules of curd have become firm, the whey is drained off and, traditionally, the curd is stacked along the sides of the vat (Figure 1.14). In this position, the granules gradually fuse together and the structure of the curd changes from soft and fragile to tough and pliable. This stage is the so-called 'cheddaring process' and, although the coalescing mass may be cut into large blocks and turned to facilitate whey drainage, nothing is done to disturb the chemical processes taking place within the curd. At the end of 1–2 hours, the curd will have acquired a structure rather like 'breast of chicken' and at this point the curd is ready for coarse milling into finger-like pieces that are small enough to encourage further whey drainage, but not so small that the developed structure is destroyed. After salting at a level of 2–3% (the figure in the finished cheese should be 1.7–1.8%, but the higher initial figure allows for losses on to equipment and/or into the whey), the curd is placed into moulds for pressing at up to 200 kN/m².

The traditional cylindrical cheese, approximately 30 cm in diameter and 14 cm in height, is then bandaged (Figure 3.2), prior to storage at 8–10 °C for several months. In North America, wax coatings may be preferred as a means of protection and the internal environment so produced does allow for normal maturation (Figure 3.3).

In more modern plants, the curd pieces are transferred, after scalding and draining, on to a slowly moving conveyor (cheddaring) system, which eventually feeds the milled curd directly into moulds or, more usually, into a block-

Figure 3.3 St George's cheese, a hard-pressed, Cheddar-like variety which is manufactured in small quantities in the Azores; controlled maturation is achieved by a wax or shrink-wrap treatment. (Courtesy of Texel, Epernon.)

former. Square blocks of cheese (18–20 kg) are produced which, on the following day, are either wrapped in film and further packed into tight-fitting plastic or wooden boxes, or vacuum-sealed into bags. This secure packaging helps to ensure a close texture in the finished cheese, as well as curtailing mould growth on the surface of the cheese. The blocks are then rapidly cooled to 5 °C, but during the maturation stage the temperature will be allowed to rise to 8–10 °C.

The period in storage transforms the raw curd into a cheese of the desired texture and flavour.

Although the nature of the associated changes is poorly understood, an experienced grader is able to monitor their progress with some degree of precision. Grading at three months is normal for all Cheddar cheese but, because maturation is a natural, biological process, mature types may have to be regraded some six months later to isolate cheeses that may not have developed along the expected lines.

In addition to guidelines for flavour and texture, typical specifications for Cheddar cheese insist that the FDM shall be 48–50%, depending upon the country of manufacture, and the moisture content shall not exceed 36–39%. In the UK, the Cheese Regulations, 1970 (SI 1970 No. 94) stipulate a minimum FDM of 48% and a maximum moisture level of 39%. Some countries also inclue an advisory specification that the salt-in-moisture (S/M) shall be in the range 4.0–6.0%, which is equivalent to 1.7–1.8% in the cheese as consumed.

The potential consumer appeal of including non-milk ingredients like chutney or wine into Cheddar cheese has been explored by some producers, but many people feel that a top quality cheese is best eaten in the traditional fashion as a component of a 'Ploughman's lunch' (Figure 3.4).

Figure 3.4 Traditional Cheddar, the best-known of all the hard-pressed English cheeses. (Courtesy of The Cheeses of England and Wales Information Bureau, UK.)

Cheshire cheese

Although Cheshire cheese is now manufactured in a number of countries, it was originally made in England in the counties of Cheshire and, to a lesser extent, Shropshire. It is one of the oldest English cheeses: according to some reports, it was produced even before the Roman legions arrived in Chester. Certainly the name appears in the 'Domesday Book', and hence its introduction must have preceded the Norman invasion by a considerable period.

The original cheese was orange in colour due, so it is said, to the inclusion of carrot juice, but nowadays it is usually creamy-white in colour; on occasions, a pale orange variant is produced by the use of annatto. Although designated as a hard cheese, it has an open, flaky texture with visible fissures of mainly mechanical origin and as a consequence the body tends to be somewhat crumbly (Figure 3.5). The mature cheese has a mild, lactic flavour, and local cheeses often acquire a characteristic 'salty' note derived from the soils of Cheshire. Most cheeses are consumed 1–3 months after manufacture, but a drier, longer maturing type is produced in certain factories.

Attainment of these characteristics is dependent partly upon the technique of cheesemaking, and partly upon the selection of the correct type of starter culture; in this case, one that includes strains of *Lact. lactis* subsp. *lactis*, together with *Lact. lactis* subsp. *cremoris*, capable of a rapid production of lactic acid. The importance of this requirement is that of shortening the time from renneting to milling, so giving a curd with the desired acidity but a higher moisture content than would be possible with slow acidification over a longer period of time. As the delicate flavour of the cheese could be spoilt by the activity of bacteria of non-starter origin, the cheese milk is pasteurized at 67–68 °C for 15 seconds prior to transfer to the vat at 30 °C (Davis, 1976). The starter culture is then added. After thorough stirring, the milk is left to ripen for about one hour. At this point, the acidity of the milk should have reached around 0.19–0.2% (as lactic acid),

Figure 3.5 Cheshire cheese, one of the oldest of the English varieties. (Courtesy of The Cheeses of England and Wales Information Bureau, UK.)

and it is then ready to receive the rennet and, if required, the annatto dye.

Some 30 minutes later, the milk will have formed a coagulum that is firm enough for cutting into small cubes of about 0.5 cm^3. Correct handling of the curd is critical, and it is important to ensure that the scalding temperature does not exceed 32–34 °C, for otherwise the curd pieces will shrink and expel too much moisture. The open texture of the curd pieces does, of course, slow down the drainage of whey from the mounds of curd that are later piled along the sides of the vat, so that while the curd does become firmer during the 'standing' stage, the close texture of a typical Cheddar curd is never achieved. Regular breakage, by hand, of the blocks of coalescing curd also restricts the development of a well-defined structure.

Once firm enough for milling (acidity 0.65–0.75%), salt is sprinkled over the curd at a rate of *circa* 2.0%, and the blocks are then finely chopped in a peg mill. Alternatively, the blocks may be milled and the salt added after the milling process, but in either case the essential aim is to ensure an even distribution throughout the bulk curd.

In the traditional process, the curd is then filled into cloth-lined, cylindrical moulds that will give a finished cheese some 30 cm in diameter and 30 cm in height. After allowing this mass of curd to drain at 20–21 °C overnight without pressure, the mould is transferred to the press. Over the next two days, the cheese, still at 20–21 °C, is exposed to increasing pressure until a value of some 100 kN/m² is reached. On completion of this stage, the mould is removed and the cheese is bandaged, surface dried, and then placed in a cool room at 10–12 °C for one week prior to transfer to the main storage facility at around 7 °C.

In more modern factories, a drier form of cheese is moulded into blocks (18–20 kg), and then wrapped in film for storage. This shape is more convenient both for mechanical handling and for cutting into consumer portions. However, the 'open' texture of Cheshire cheese makes it susceptible to the growth of moulds, so that correct sealing of the plastic film is a critical operation.

Grading of the finished cheese is undertaken at an early stage, and although the sensory characteristics are the major concern, the retail cheese must meet certain legal requirements. The relevant regulations (UK) declare that Cheshire cheese must have a minimum FDM of 48% – equivalent to around 27% in the cheese as consumed – and that the moisture content must not exceed 44%.

Derby

Derby is a little-known cheese with a composition close to Red Leicester (described later) in that it must contain at least 48% FDM and have a maximum moisture content of 42%. The manu-facturing procedure is not unlike that for Cheddar, but because the final texture is softer than for Cheddar, the handling of the curd is more gentle. Pasteurized cow's milk is the usual base material and, after warming to 28–30 °C, the milk is inoculated with an appropriate starter culture. As the acidity develops, rennet is added at the rate of around 30 ml/100 litres of milk and the bulk is thoroughly stirred to distribute the enzyme. After 45–50 minutes, the coagulum is ready for cutting with knives with gaps of at least 1 cm. Gentle stirring follows and over the next 50–60 minutes the temperature is raised to 36 °C. This gentle heat treatment gives curd pieces that are firm, but not hard like Cheddar curds, and these are allowed to settle to the bottom of the vat. Once a mild acidity of around 0.21% lactic acid has been reached (Davis, 1976), the whey is drained off to leave a semi-solid mat of curd. Because this mass is cut into blocks, which are piled along the sides of the vat (Figure 1.14), the cheese receives a texturizing stage typical of this group of cheeses. Acid development continues during this time, and at 0.7% lactic acid the curd is milled and salted at a level of 2% or there-abouts.

As a variety that is not manufactured for the mass market, many cheeses are still produced in the traditional round moulds, with the finished cheese being bandaged in calico for storage. After pressing, which traditionally involved turning the cheese at regular intervals and gradually increasing the pressure, the cheeses are stored at an even temperature of 14 °C for at least one month, and more usually 3–4 months to acquire the desired flavour and texture; the latter is characteristically 'flaky' compared with Cheddar and somewhat softer. Film-wrapped blocks are available also for ease of retail handling.

Although distinctive in detail from the other territorial cheeses, Derby has suffered to some extent from being a cheese for the connoisseur and, in an attempt to provide a more obvious identity, Sage Derby was created (Figure 3.6). It is manufactured by moistening ground sage leaves with a solution of chlorophyll extracted from spinach, kale or some other suitable source, and

Figure 3.6 The green, mottled appearance that is so characteristic of English Sage Derby. (Courtesy of The Cheeses of England and Wales Information Bureau, UK.)

then adding the mixture to the curd. Sometimes white and coloured curds are employed to give a layered appearance, but the overall marbled effect can be equally pleasing. The intensity of flavour can be adjusted according to market demand, but the herb should not be allowed to become too dominant.

Gloucester

Gloucester, as with other territorial cheeses from England and Wales, derives its name from the county of origin. It was originally produced in two forms: Single Gloucester and Double Gloucester. Both were about 40 cm in diameter, but while Double Gloucester was 15–20 cm in thickness (Figure 3.7), Single Gloucester would have been some 6–8 cm. It is reported that the Single form was mainly a true farmhouse cheese in the sense of being made by the farmer's wife in any available outhouse, and whether the lower weight of 6–7 kg was selected as being more convenient for handling is a matter for conjecture. Irrespective of size, Gloucester is a firm, close-textured cheese with a mild, rich flavour and slightly waxy appearance; a low level of annatto is often added to give the cheese a distinctly yellow–orange colour.

As with many territorial cheeses, the procedure has much in common with that for Cheddar and, as discussed earlier, these similarities are probably a reflection of the constraints introduced by the climate. Subtle differences in the methods of

Figure 3.7 Double Gloucester, one of the classic English cheeses that is not widely manufactured. (Courtesy of The Cheeses of England and Wales Information Bureau, UK.)

handling are important, and the manufacture of Gloucester is a good illustration of this point.

The milk – pasteurized and standardized cow's milk – is first ripened for around one hour with a mesophilic starter culture, and then sufficient rennet is added to provide a firm coagulum some 60 minutes later. The gel is gently cut into pieces about 5 mm along the sides, and over the next hour the temperature of the vat is raised to 35–37 °C. A further period of stirring at this temperature produces a firm curd from which the whey is slowly drained, over a period of perhaps 30 minutes (Davis, 1976). A weighted rack is used, as for Leicester, to express more whey and compress the curd ready for cutting into blocks (15 cm). The subsequent handling of the cheese is again not unlike Leicester, except that the curd will be milled twice and the acidity will be around 0.8% as lactic acid as against 0.60–0.64% for Leicester. About 2% salt is usual for Gloucester, and the moulds are selected nowadays to give a traditional cheese with a diameter of 35 cm and a height of 22 cm. However, as with many cheeses, the trend is towards forms that are easier to handle in-store and cut for retail portions (Tamime, 1993).

The Cheese Regulations state that Double Gloucester should have a minimum FDM of 48% and a maximum water content of 44%. The higher moisture level than Cheddar tends to encourage more rapid maturation, and most cheeses are ready for consumption at 4–6 months. Single Gloucester, although rarely made for the mass market, tends to have a marginally higher moisture content and shorter shelf-life than the Double form, as does a similar cheese – Cotswold. According to Davis (1976), Cotswold cheese is all but identical to Single Gloucester, and is clearly a local variant.

Friesian Clove and Leiden

Friesian Clove (Figure 3.8) and Leiden (Figure 3.9) are two firm, dry-textured cheeses associated with the province of Friesland and the town of Leiden (Holland), respectively.

They may be made from either full-cream or skimmed milk, and a small volume of buttermilk is added by some farmhouse makers to enrich the flavour. Mesophilic starter bacteria are employed so that, after ripening and renneting at around 30 °C, the cut curd is not scalded much above 35 °C. After the whey has been removed, the curd is collected (often into cloth sheets) and the herbs are blended and kneaded into the cooling mass. Caraway, cumin and cloves are the usual choices, and the precise blends vary from producer to producer. Originally, only a portion of the curd was flavoured, so that the finished cheese would appear as a sandwich of natural curd with a 'filling' of spiced cheese. Nowadays, it is usual for all the curd to be spiced, and then transferred to cloth-lined moulds for pressing. An overnight treatment, perhaps with an inversion within the press after 3–4 hours, is sufficient to produce the desired firmness, and the cheeses are then ready for dry salting or immersion in brine.

Curing in a cool, moist store for a few months allows time for the flavour of the herbs to penetrate the curd. To prevent the rind from becoming too hard, the cheeses may have to be wiped with brine from time to time. A pale yellow colour and compact texture are characteristic of both cheeses, and some typical analyses are shown in Table 3.1. The major difference between the figures for each cheese is the fat content, and the values of 20% or 40% FDM are the accepted standards for the retail products. In terms of chemical composition, the similarity between the two varieties is evident.

Table 3.1 Typical compositions (%) of Friesian Clove and Leiden

	Friesian Clove	Leiden
Protein	39	33
Fat	13	27
Fat-in-dry-matter	20	40
Moisture	42	34
Salt	2.9	2.5

Figure 3.8 Friesian Clove cheese (8 kg), a speciality of factories in the province of Friesland, is a dry, firm-textured cheese with a flavour derived from the added cloves and cumin. (Courtesy of the Dutch Dairy Bureau, Leatherhead, Surrey, UK.)

Figure 3.9 Leiden cheese, originally a farmhouse variety produced around the town of the same name, has a firm, dryish texture, and the piquant flavour comes from the inclusion of cumin seeds. (Courtesy of the Dutch Dairy Bureau, Leatherhead, Surrey, UK.)

Greek Graviera

Graviera is a comparatively recent addition to the cheeses of the Mediterranean, and its method of production began to achieve a degree of standardization around 1920 (Anifantakis, 1991). Even so, regional differences are still evident, and the Graviera from Crete (Figure 3.10) would have a flavour that reflected the geography of the island.

It is made largely from sheep's milk, but cow's milk or mixtures of milks are used by some producers. Raw milk is still favoured by some small producers, but most Graviera is produced in factories from pasteurized milk inoculated with around 1.0% of a culture of *Lac. lactis* subspp. *lactis* and *cremoris*, together with 0.1% of a culture containing a mixture of *Str. thermophilus* and *Lac. helveticus*; 0.1% of a yoghurt culture may be used as an alternative to the latter pairing. Calcium chloride (10 g/100 litres of milk) is added to compensate for any deficiencies in calcium ion content, along with sufficient standard rennet to achieve coagulation in 30–40 minutes at 35 °C.

Cutting into cubes with sides of 5 mm follows and, after a short period at rest, stirring begins. A steady increase in the temperature of the curd/whey mixture to around 50 °C gradually increases the firmness of the curd particles, which are then allowed to settle to the bottom of the vat. In small dairy plants, cloth-lined moulds are prepared to receive the curd and, over the next few hours, the moulds are subjected to varying degrees of pressure and the cloths changed at regular intervals. Next day, the cheeses are removed from the moulds and dry salted, and a routine of turning and dry salting is maintained for a further 2–3 weeks; in all, each cheese may receive 40–50 additions of dry, coarse salt (Anifantakis, 1991). An additional maturation period of 3–4 months follows, during which time the cheeses are turned and wiped at regular intervals with a brine-soaked cloth.

In modern factories, the curd handling and pressing stages are automated for convenience, and immersion of the cheeses in a brine bath

Figure 3.10 Greek Graviera, a hard cheese made from cow's or sheep's milk, or from mixtures of any available milks. The precise origin of the cheese (e.g. Crete or Naxos) usually forms part of the trade name. (Courtesy of Professor E.M. Anifantakis.)

reduces the need for dry salting during the early days of maturation; there is still a need for 20–25 additions of dry salt during the later period of storage.

The typical composition of Graviera might be as follows:

Protein	25%
Fat	37%
Moisture	34%
Salt	1.5%

On cutting, the cheese is firm and slightly elastic, and natural fissures are common. In some samples, round 'eyes' are clearly visible as well; these arise mainly when raw milk has been employed, and *Propionibacterium* sp. has produced carbon dioxide during the maturation stage. Some factory-made products will also have 'eyes' if the bacterium has been added along with the starter culture, but the presence of 'eyes' tends to be an optional rather than an essential feature of Graviera.

Idiazabal

Although the annual production of this cheese is only around 350 tonnes, it is a variety officially recognized in Spain. It is made in the mountains of the Basque country from sheep's milk, and is matured in natural caves in the same region.

The milk is heated in a convenient vat to 38 °C, and coagulated in 10–15 minutes with animal rennet. After allowing the gel to cool to 25 °C, and presumably the natural bacterial flora to develop, the coagulum is broken up and filled into wooden moulds. Once the curd has consolidated, the cheeses are dry salted or immersed in brine. The finished cheeses (Figure 3.11) are cylindrical (16–25 cm in diameter and 8–12 cm in height), and they are transported to the mountains for storage at ambient temperature.

Two months later, the cheeses will be removed from the caves for smoking in kilns fired with beech wood, and then returned for a further period of up to one year. The mature cheese is firm with only mechanical fissures in evidence, and is white–yellow in appearance; the natural flavour of the cheese is enhanced by a hint of 'wood smoke'.

A typical composition for Idiazabal might be:

Protein	23%
Fat	38%
Fat-in-dry-matter	56%
Moisture	33%

Figure 3.11 Idiazabal, a hard cheese made from sheep's milk in the Basque country and also Navarre. (Courtesy of the Ministerio de Agricultura Pesca y Alimentacion, Spain.)

Kefalotiri

Kefalotiri is a traditional Greek cheese with a hard texture and strong, salty flavour. The name, as with Ras cheese, is supposed to refer to the 'head-shaped' appearance of the cheese (Figure 3.12), but 'kefalos' in Greek can also imply 'hat-shaped', so the true origin of the name remains a matter for speculation. According to Anifantakis (1991), around 4500 tonnes of this cheese are consumed annually in Greece, with the availability of sheep's or goat's milk being one of the factors that determines supply. Consequently some European countries, and to a lesser extent Greece itself, have introduced a Kefalotiri cheese made from cow's milk. However, whether the latter product achieves the organoleptic quality of the traditional cheese, or indeed should be called Kefalotiri at all, is contested by some authorities.

Full-cream milk, or milk standardized to around 6.0% fat, provides the base material and will be used in the raw state if the microbiological quality is excellent. The flavour of cheese made from raw milk is usually rated as superior to cheese made from pasteurized milk but it has become the standard practice to heat-treat the milk at 68 °C for around 10 minutes or 72 °C for 15 seconds, depending upon the equipment available. This trend reflects both the desire to avoid any risk from pathogens, and the fact that milk of indifferent microbial quality can give rise to cheeses which spoil during ripening. After cooling the milk to around 35 °C, a starter of culture consisting of *Str. thermophilus* and *Lac. bulgaricus* may be added at a rate of around 1%, although *Lac. casei* rather than *Lac. bulgaricus* has been advocated by some workers (Zerfiridis and Pappas, 1989). By adding the culture at least 30 minutes in advance of renneting, acid development is well established by the time the rennet is added along with calcium chloride (10 g/100 litres of milk).

The calcium chloride is added to compensate for any deficiency of calcium ions in the base milk, for low levels can both delay coagulation and detract from the quality of the coagulum.

Figure 3.12 Kefalotiri, a hard cheese made in Greece from sheep's milk. (Courtesy of Professor E.M. Anifantakis.)

The rennet itself may be the standard commercial form, or it may be a crude local extract containing a mixture of enzymes, including lipases; the presence of this latter group gives rise to a cheese with a pleasant, piquant flavour. A period of around 30 minutes at 35 °C ensures a coagulum of sufficient firmness for cutting into particles of up to 1.0 cm^3. While the contents of the vat are being stirred, the temperature is slowly raised to 43–45 °C, and then held at this point for a further 15 minutes or so. Once the curd has settled, some of the whey is drawn off and the curd transferred into moulds lined with cheesecloth. Manual pressure is employed to expel the free whey, and once the cheese has begun to take shape the cheesecloth is removed and the cheese reversed in the mould along with a clean cloth (Anifantakis, 1991). Mechanical pressure is then applied for a short period, followed by a change of cloth and reversal once more in the mould. After this change, the pressure is slowly built up to reach around 10 times the weight of the cheese, with breaks at hourly intervals to reverse the cheese in its mould.

After some 4–5 hours, the cheesecloth is removed; the cheese is returned to its mould and placed in an overnight store at around 15 °C. Next day, the cheeses are placed in a brine bath. Although some salt will be absorbed during this stage, the final salt content of around 4% is achieved by dry salting. This stage involves rubbing the exposed surfaces of the cheese every day (or alternate days) with dry salt for around three weeks, along with occasional washes in brine to avoid the risk of microbial growth. One final wash with brine precedes transfer to a cold store at <4 °C and a relative humidity of 85%. Ripening for around three months overall gives a cheese with a moisture content of 36%, a fat content of 28% (45% FDM) and a pH of around 5.0, but variations between the different regions of origin are widespread.

Leicester (Red Leicester)

As the name implies, Leicester was originally made in just one county of England and, even today, it is not manufactured widely. It is not unlike Cheshire in many respects, but has a more intense colour and higher moisture content. It is usually quite a small cheese (Figure 3.13) being only 35–40 cm in diameter and 15 cm deep, but the soft, pliable texture, mild flavour and distinctive bright red colour make it a popular addition to the cheese board.

Originally, a mixture of raw morning and evening milks coloured with annatto (45 ml/100 litres of milk) was warmed to a temperature of 25–27 °C, with the precise figure depending upon the season; starter (1% v/v liquid or equivalent) and rennet were then added. About 80–90 minutes later when a strong coagulum had formed, the gel was cut into pieces using 1 cm cheese knives. Careful stirring for about 10 minutes was followed by scalding at 33 °C for around one hour. The whey was then drained off, and weighted racks placed on to the curd further compressed the solid mass. The partly consolidated material was then cut into blocks of some 18–20 cm and piled on a draining rack; a cloth was often draped over the curd to prevent cooling and hindrance of microbial activity. Sometimes

Figure 3.13 Red Leicester, a hard-pressed English cheese that is not widely manufactured. (Courtesy of The Cheeses of England and Wales Information Bureau, UK.)

the blocks would have been cut once more and inverted to assist whey drainage, and once the acidity and firmness had reached the desired level, the curd was milled and salted (2%). Moulding and pressing with a gradually increasing pressure followed. The finished cheeses were then lightly dry salted and allowed to mature at around 15 °C for 6–8 months; some cheeses, however, were suitable for consumption earlier.

Nowadays, good quality pasteurized milk forms the base material, and the ripening temperature tends to be standardized at 30 °C. After ripening for 45–60 minutes, sufficient rennet is added to give a coagulum some 40 minutes later. Cutting is normally judged to provide pieces of a smaller size than previously, and it is then scalded at 37 °C for around 40 minutes. At the end of this time the pieces will be quite firm, and the curd is then held for a further period until the whey is drained off. A weighted rack or board may be used to speed up the loss of whey, which by this time may have reached an acidity of 0.3–0.4%. Cutting of the curd mass into blocks tends to follow the traditional pattern, as does the milling, moulding and pressing. However, blocks will replace rounds where the process plant allows, and vacuum-packing may be employed to discourage the growth of moulds on the exposed surfaces rather than dry-salting.

The UK regulations demand a final FDM of 48%, and moisture content not above 42%, but beyond that the characteristics of the cheese depend on the skill of the cheesemaker.

Manchego

Manchego is perhaps one of the best-known cheeses outside the Iberian Peninsula, even though only around 1600 tonnes of true Manchego (i.e. coming under the Appellation d'Origine) are produced each year. As a consequence, an increasing number of similar cheeses made from cow's or goat's milk are appearing in the market-place.

However, Manchego retains its unique character through the use of sheep's milk. Raw or pasteurized milk may be employed, but factory production is usually based upon milk heated to 72 °C for 15 seconds. After adjustment of the temperature to 30–32 °C, calcium chloride and a mesophilic starter culture are added. The latter is a mixture of homofermentative species, i.e. *Lac. lactis* subspp. *lactis* and *cremoris*, and heterofermentative types such as *Lac. lactis* biovar *diacetylactis* and *Leuconostoc mesenteroides* subsp. *cremoris*. Under farmhouse conditions, the range of species will be more extensive again but, in any event, the mixed flora does help to provide Manchego with a complex flavour profile.

Once the starter culture has produced a detectable level of acidity – perhaps up to 0.25% lactic acid – sufficient standard rennet is added to give a firm coagulum in around 45 minutes. The gel is then cut into pieces with sides of about 5 mm, and the temperature of the vat is raised to 38 °C. Stirring at this temperature for 15–20 minutes imparts a degree of firmness to the curd, and it is then transferred to metal moulds that will give

rise to cheeses of 18–22 cm diameter and 8–10 cm thick. The cheeses are pressed for 12–16 hours, perhaps with turning at some intermediate point, and then finally bound in basket-work wrappings to give the characteristic markings to the sides of the cheese (Figure 3.14). After 30 minutes or so, the bindings are removed and the cheeses placed in a bath of strong brine (24–26% salt) for 24–48 hours. Final maturation, for at least two months, takes place in a store at 10–15 °C and 80% RH, and the cheeses are turned frequently; the surfaces of the cheeses are often cleaned with olive oil to improve the appearance of the rind. A typical cheese might weigh 2–3.5 kg.

Figure 3.14 Manchego, a hard to semi-hard cheese made from sheep's milk in the La Mancha region of central Spain. (Courtesy of the Ministerio de Agricultura Pesca y Alimentacion, Spain.)

The composition might be:

Protein   23%
Fat   34%
Fat-in-dry-matter   54%
Moisture   37%

The flavour of the mature cheese is strong for many palates – a feature that is enhanced by the practice of some producers of adding the enzyme, lipase, along with the rennet – and for this reason some Manchego is sold inside the two-month maturation period. If this expedient is followed, the cheese should be clearly labelled, e.g. Manchego fresco.

Ras cheese

Ras cheese is an Egyptian hard cheese that is rather similar to the Greek cheese, Kefalotiri. The names from both countries mean 'head' – perhaps because the cheese resembles a bald head (Figure 3.15) – and it is probable that the basic cheese originated in the Balkans. However, Ras cheese is now the best-known hard cheese in Egypt, and indeed throughout the Arab world, and the best quality cheese is reported to be made by private cheese companies in the Damietta Governorate of Egypt.

Standardized milk (3% fat) is heated to 32 °C, and sufficient rennet is added to complete coagulation in 35 minutes (Abou-Donia, personal communication). The coagulum is cut into small pieces, about the size of wheat grains, and then vigorously stirred. The temperature of the vat is then raised to 45 °C over a period of around 40 minutes, and gentle stirring is maintained throughout. After the curd has settled and the whey drained out, salt is sprinkled over the curd at a rate of 1% (w/w), and the curd is manually pushed to the sides of the vat. Moulds, lined with cheesecloth, are filled with sufficient curd to produce one finished cheese, and manual pressure is applied to expel some of the adhering whey. Light mechanical pressure follows over the next 4 hours, at which point the cheese is reversed in the press and left under pressure for 24 hours.

Figure 3.15 Ras, a hard cheese produced in various regions of Egypt. (Courtesy of Professor S.A. Abou-Donia.)

The wheels of cheese are then removed from the moulds and cloths and placed in brine (20% salt solution) for 24 hours. After draining for a further day at ambient temperature, the surfaces of each cheese are coated with a small quantity of dry salt. By the following day, most of this salt will have been absorbed into the cheese, so that the wheels are turned and the dry salting process repeated once again. This dry salting procedure is continued for a period of around two months, either daily or every other day. If a cheese becomes hard, wet salt may be used instead, but this problem is usually avoided by washing the cheeses in brine at least twice a week.

The fresh cheese should have a moisture content below 39% and an FDM level of at least 50% but, as maturation proceeds, the values may change to 30% and 55%, respectively (Hofi *et al.*, 1970). When young, Ras cheese is not unlike Edam in texture, but the presence of 'eyes' tends to depend on fortuitous contamination rather than design.

Roncal

This cheese derives its name from the Roncal Valley in northern Spain, and 400 tonnes is produced each year under the Appellation d'Origine.

Like many cheeses in the area, it is made from raw sheep's milk and hence the active microflora is rich and varied. However, in spite of this mixed population of bacteria, the rate of acidification and ripening of the milk tends to be slow, and coagulation by standard rennet may take over 60 minutes at 37–40 °C. After being cut into small pieces, the curd is allowed to settle to the bottom of the vat, and the whey is slowly drained off over the next 16–24 hours. Manual 'pressing' against the sides of the vat may assist drainage, so that blocks of matted curd can then be broken by hand and placed in moulds. The application of light pressure helps to expel the whey from the curd. At one time the cheeses were removed from the moulds after 24–48 hours and air-dried to form an outer coat. More usually, a period of light pressure for 3–4 days is sufficient to produce the final form of the cheese – 18 cm in diameter and 9 cm thick – and it is then dry salted for around a week.

The cheeses are smoked with sufficient intensity to give them the brown, leathery coat shown in

Figure 3.16 Roncal, with its origins in the valley of Roncal, Navarre, is similar in many respects to Manchego cheese. (Courtesy of the Ministerio de Agricultura Pesca y Alimentacion, Spain.)

Figure 3.16, and then placed in caves or storerooms at 6–8 °C and 100% RH for maturation over 45–50 days. Regular turning of the cheeses helps to ensure that flavour development is even and regular. The end result is a hard cheese with mechanical openings in the white or pale yellow body rather than 'eyes', and a strong, slightly sharp flavour. The final weight is usually 1.8–2.0 kg.

A typical composition might be:

Protein	25%
Fat	39%
Fat-in-dry-matter	55%
Moisture	29%

Serena

Serena is made in the Valley of La Serena in western Spain, and the milk of Merino sheep is the raw material. It is a flat, cylindrical cheese (Figure 3.17) about 17–20 cm in diameter and 7–9 cm thick, and weighing in the region of 1.0–1.5 kg. It is made by warming the milk in a vat to 32–36 °C, and then adding a vegetable coagulant/extract obtained from the cardoon, *Cynara cardunculus*. Slow coagulation over the next 60–90 minutes gives rise to a gel that can be cut into cubes with sides of up to 2 cm. Light pressing allows the finished cheese to retain a somewhat open texture, but dry salting and maturation at ambient temperature and humidity act in concert, over a period of 1–2 months, to produce a firm-bodied cheese (moisture content 39–42%) that can withstand storage for over a year. The FDM value is in the region of 50–52%.

Figure 3.17 La Serena cheese derives its name from the valley in western Spain where it is produced from sheep's milk. (Courtesy of the Ministerio de Agricultura Pesca y Alimentacion, Spain.)

4 Cheese varieties designated as 'semi-hard'

The description of a cheese as semi-hard or semi-soft is largely subjective, but the phrases do convey an essential difference in character. Thus, whilst a semi-soft cheese might be expected to crumble so easily as to almost reach the point of 'spreadable', a semi-hard cheese is usually 'sliceable', albeit into coarse sections. Consequently, the division has been employed to bring together those cheeses which are not 'hard' in the sense of Cheddar, but which will still readily portion for retailing as individual blocks.

Barra

This cheese is much like an Edam (see later), in that it has a very close, elastic texture (Figure 4.1) and is capable of being sliced with ease. The precise square shape is relevant also, in that it has the correct dimensions for cutting for use in hamburgers or sandwiches. It is produced in Argentina in some quantity, and is matured for one month before vacuum-packing into plastic bags and distribution, mainly to the catering trade. Household consumption is often with jam or marmalade and, because it is never eaten as the principal component of a snack, the flavour of the product tends to be regarded as of secondary importance to texture.

Liplovska Bryndza

The name Bryndza or some similar name occurs quite widely in the literature on cheeses, but this traditional cheese from Slovakia is unique, not least in its method of manufacture.

Raw sheep's milk with 6–7% fat is the base material. After adjustment of the temperature to around 30 °C, rennet is added along with calcium chloride. The choice of rennet will vary from producer to producer, for while some favour the standard material of commerce, others prefer a crude extract that contains more lipase. Coagulation takes place in 40–45 minutes, and the curd is then cut into small pieces of some 5–15 mm along the sides. On standing, the curd settles to the bottom of the vat, and once the whey has become clear, the major portion of the liquid phase is removed. The partially exposed curd is then pressed together by hand to form 'lumps' of 4–5 kg, which are removed as soon as sufficiently firm and placed into cloth bags. By hanging the bags, initially over the vat and later in a ripening room at 20 °C, further drainage of the whey occurs as well as a rise in acidity. After some 24 hours, the 'lump' of curd is sufficiently firm to be removed and stood on a shelf in the same room. Over the next three days, the native microflora – including a range of streptococci and lactobacilli – continues to secrete lactic acid, and probably other acids like acetic acid as well, until the detectable acidity has reached in excess of 2.0%. This acidity signals the end-point of the first stage of manufacture, and this intermediate product is known locally as Hrudka cheese.

The irregular masses of Hrudka cheese are then shipped to a central factory for the manufacture of Bryndza. At the factory, the Hrudka cheeses, which may have come from numerous farms and small dairies in the vicinity, are graded according to quality. The soft rind is then removed from the

Figure 4.1 Barra, a close-textured cheese of sufficient elasticity to allow easy slicing for sandwiches or other fast-food uses. (Courtesy of Dr Carlos Zalazar, Instituto de Lactologia Industrial, Argentina.)

selected cheeses, which are broken into pieces, mixed with around 2.0% salt, and fed between granite rollers to give a smooth, homogeneous paste (Scott, 1986). Although smaller containers are used in some factories, polythene-lined beech-wood tubs (Figure 4.2) are the preferred method of packing and storage. After a maturation period under cool conditions, the end-product is a white, firm cheese – perhaps erring on the side of semi-soft – with a mild, slightly acid flavour.

Figure 4.2 Liptovska Bryndza, a traditional Slovak cheese made from sheep's milk on small farms in the mountains. (Courtesy of the Dairy Research Institute, Zilina.)

Caciotta

Italy has a long tradition of farmhouse cheese-making, and Caciotta is a general name for small, semi-hard cheeses made throughout the country. Sometimes the cheese will be given a local name, such as Fresa in Sardinia (Davis, 1976), but they are all mild, quick-ripening cheeses intended for table use.

The milk of cows, sheep or goats is used according to availability, and a hint of saffron may be added to give the finished cheese a distinctly yellow colour. After warming to 35 °C, standard rennet is added to the milk and, if a softer cheese is desired, use may be made of a vegetable extract from the cardoon (*Cynara cardunculus*). Some 30 minutes later, the curd will be ladled into open-ended, cloth-lined moulds standing on a draining mat. After natural draining for 24 hours or more, the cheeses are firm enough to be removed from the moulds and dry salted.

A wide variety of size and finishes is available (Figure 4.3) but Caciottas are usually ready for eating in 10–20 days.

A typical composition might be:

Protein	23%
Fat	29%
Fat-in-dry-matter	45%
Moisture	40%

Figure 4.3 Caciotta, a medium-hard cheese produced in central Italy from either cow's or ewe's milk or a mixture of both. (Courtesy of the Ministero dell'Agricoltura e delle Foreste, Rome.)

Caerphilly

Caerphilly is a white, crumbly cheese with a mild, acid flavour, and it was first produced in the town of the same name in Wales. It is an attractive option for small producers in that it matures in 10–14 days, and hence the need for costly storage facilities can be kept to a minimum.

Most production is now factory-based, and full-cream milk is first pasteurized at 69–70 °C for 15 seconds, and then cooled to 32 °C. An active homofermentative culture of *Lact. lactis* subspp. *lactis* and *cremoris* is added, and when the acidity of the milk has reached 0.2% lactic acid or thereabouts, sufficient standard rennet is added (20–25 ml/100 litres of milk) to form a firm gel in around 45 minutes. The coagulum is then cut into cubes with sides of 6 mm, and gently stirred for about 15 minutes before the temperature is gently raised by 2 or 3 °C. Exposure to this gentle scald is enough to produce a firm curd, and over the next hour the pieces of curd are allowed to settle and the whey is run off. The curd is then piled along the sides of the vat and/or into mounds. Some texture formation does occur during this time, but frequent disturbances of the 'lumps' of curd tends to prevent the pieces coalescing to any degree.

However, the quite rapid production of lactic acid does continue (few starter bacteria are killed by the low scalding temperature) and at around 0.24% total acidity, salt is mixed into the curd at a rate of 1.0%. The salted curd is then transferred to cloth-lined moulds and, after pressing for one hour, the cheeses are turned and subjected to further pressure. Around three hours later, the cheeses are removed from the press and dry salted, prior to being placed back in the press and subjected to a final pressure of 49 kN/m^2. Next morning, the cheeses are removed from the moulds and cloths, and placed in a brine bath (18% salt) at 15–16 °C. Twenty-four hours later, the salt content of the cheeses will have stabilized in the region of 2.0%, and the cheeses are dried with clean cloths. A short period of maturation at 10–15 °C follows.

Traditionally, each cheese would have been

Figure 4.4 Caerphilly, a cheese with its origins in Welsh farmlands. (Courtesy of The Cheeses of England and Wales Information Bureau, UK.)

dusted with rye flour, barley meal or some similar material to give a dry coat (Figure 4.4), but nowadays film wraps are proving more popular. The precise dimensions of the cheese tend to vary quite widely, but an overall weight of 3–4 kg would be typical. The current cheese regulations demand a minimum FDM content for Caerphilly of 48% and a moisture level of below 46%.

Colby

Although Colby cheese (Figure 4.5) is not unlike Cheddar cheese in many respects, it tends to contain more moisture than Cheddar and hence the texture is rather softer. The initial stages of manufacture are also broadly similar, but the later processing steps are characteristic of a number of other cheeses from North America, in

Figure 4.5 Colby, originating in Colby, Wisconsin, a hard to semi-hard cheese with a texture not unlike Cheddar. (Courtesy of the Wisconsin Milk Marketing Board.)

that the curd is vigorously stirred after whey drainage.

In practice, pasteurized full-cream cow's milk is put into a vat at 31–32 °C, and an active mesophilic starter is added at a rate of around 1.0%. After some 30 minutes, standard rennet (20 ml/100 litres of milk) is added, and a firm coagulum should be produced in 30–35 minutes. The gel is then cut and stirred in the normal way, followed by scalding to 37 °C and holding for 45 minutes at the elevated temperature (Kosikowski, 1982). As the firm pieces of curd settle to the bottom of the vat at the end of scalding, most of the whey is drained off and then replaced with clean, cold water until the temperature of the whey/curd mass is 25–26 °C. The suspended curd is stirred for around 15 minutes, and then most of the whey is removed. Vigorous stirring of the concentrated mass follows, and some 10 minutes later the remainder of the dilute whey is removed. The dry curd is then stirred for further 20 minutes so that, unlike the situation with hard-pressed cheeses, no structure is allowed to develop in the vat. Coarse salt is usually added to the curd

during the dry-stirring stage, and the aim is to produce a finished cheese with a salt concentration of 1.7%.

The granular curd is then transferred to moulds for pressing, initially quite gently, and later overnight at a higher pressure. Next morning, the cheeses are placed in a curing room at 4–5 °C and left, with occasional turning or cleaning, to mature for 2–3 months. It is a faster ripening cheese than a typical hard cheese and this contrast is a reflection, in the main, of the higher moisture level – typically around 38–39% for Colby as against 33–34% for Cheddar or 31–32% for Gruyère. The FDM should be in excess of 50% (USDA, 1974).

The rindless block shown in Figure 4.6 is a popular variant of Colby made by mixing curds coloured with annatto with the natural-coloured material. The marbled effect is really quite striking and makes the cheese a popular purchase for parties or other social gatherings.

A broadly similar cheese from North America is Monterey, which differs from Colby in that the full-cream form tends to be a little softer in texture. There is also a grating variant manufactured from skimmed or semi-skimmed milk, and this latter type is matured for around six months as against six weeks for the table cheese.

Figure 4.6 Colby Jack, a distinctive, rindless variant of the traditional cheese. (Courtesy of the Wisconsin Milk Marketing Board.)

Lancashire

Lancashire cheese was made originally only in the county of the same name in England and, as a quick-ripening variety, it tended to be consumed locally. It is a demanding cheese to manufacture and this aspect may have contributed also to the reluctance of cheesemakers elsewhere to copy the technique.

Basically, a curd is manufactured in the normal manner from pasteurized milk. After being cut into quite small pieces, the curd is stirred at 30 °C but not scalded. The whey is then drawn off and the coalescing curd is cut into blocks prior to transfer to cloths held in a drainer. Additional breakage of the curd over the next hour encourages further whey drainage. When the curd has become firm enough (acidity around 0.2%), the curd is tipped into storage vats at ambient temperature (20–22 °C). Some 24 hours later, the acidity of the curd will have climbed to 1.2–1.4% lactic acid, and it is at a suitable stage for incorporation into the finished cheese.

This next step involves mixing the old curd with a batch of fresh curd (0.2% lactic acid) to give a blend with a final acidity of 0.9% lactic acid (Scott, 1986), and then milling the mixed curds to ensure an even distribution. Around 2.0% salt is added, and composite curd is packed into cloth-lined moulds. There is some natural drainage overnight, but increasing mechanical pressure over the next three days is mainly responsible for the final texture. Storage temperatures between 12–18 °C allow for maturation in three weeks at one extreme or up to four months. The finished cheese (Figure 4.7) has a loose, open texture and a mild, slightly acidic flavour. The body of the product is often described as 'silky' and, although firm, it is sufficiently soft in some instances to be almost spreadable.

A typical analysis might be:

Fat	23–25%
Fat-in-dry-matter	48% (min.)
Moisture	43–45% (48% max.)
Salt	1.5%

Figure 4.7 Lancashire, a semi-hard variety of English cheese that is demanding to make. (Courtesy of The Cheeses of England and Wales Information Bureau, UK.)

Mahon

This cheese derives its name from the town of Mahon on the island of Minorca, and about 3000 tonnes are produced annually. In 1980, the cheese was recognized as a distinct variety and awarded an Appellation d'Origine and, as a result of the more stringent regulations, only around 60% of the annual production now reaches the market place; the remaining output is sold for the production of processed cheese.

Figure 4.8 Mahon cheese, a quick-ripening, semi-hard variety that derives its name from the town in Minorca. (Courtesy of the Ministerio de Agricultura Pesca y Alimentacion, Spain.)

It is made from raw cow's milk which is warmed to 30 °C, and then coagulated with rennet. The natural microflora of the milk, predominantly *Lact. lactis*, *Lac. plantarum*, *Lac. casei* and *Str. durans*, provides enough acidity for the coagulant to form a gel in around 45–50 minutes. Once the coagulum has formed, it is cut into pieces with dimensions of approximately 0.5 cm. Curd, sufficient to give a finished cheese of 2–4 kg (Figure 4.8), is then collected into a cloth and pressed. Immersion in a brine bath for around 48 hours provides the cheeses with their desired salt content, and they are then dried and stored at 18 °C for three weeks.

A typical chemical composition of Mahon might be:

Protein	27%
Fat	33%
Fat-in-dry-matter	48%
Moisture	32%

It is unusual to find a semi-hard cheese with such a low moisture content (Marcos, 1987). Whether or not excessive proteolysis by the native microflora is responsible for the soft texture is not clear, but certainly it is an unexpected feature of Mahon, as is the high salt content.

Majorero

Although only a few hundred tonnes of Majorero are produced each year on Fuerteventura Island, it is a typical example of the rather strongly flavoured cheeses that can be derived from goat's milk. The flavour is derived in part from the natural microflora of lactobacilli, lactococci, streptococci and leuconostocs introduced by both the raw milk and the utensils, and from the use of rennet paste for coagulation. This latter material is obtained by macerating the stomachs of kids that have been fed only on milk. On removal, the stomachs – filled with curdled milk – are sun-dried for at least two months prior to use.

As production is essentially farm-based, the rennet paste is added to the milk at ambient temperature (20–22 °C) and the contents of the vat are left to coagulate. The gel is then cut by hand and transferred to moulds of braided palm leaves (hence the pattern visible in Figure 4.9), drained and lightly pressed. Dry salting follows, and maturation at ambient temperature allows the cheeses to surface-dry to give a soft, pale brown rind.

The typical composition of Majorero cheese is:

Protein	25%
Fat	23%
Fat-in-dry-matter	43%
Moisture	45%

Monterey

As mentioned earlier, Monterey cheese – or Stirred Curd Cheddar, as it is sometimes called –

has much in common with Colby, but it is none the less a well-known variety in its own right. Originally a farmhouse cheese produced in Monterey County, California, it is now manufactured widely in two forms: whole-milk Monterey for table use (Figure 4.10) and a skimmed-milk variant that is much harder and used for grating.

According to Kosikowski (1982), pasteurized milk at 32 °C is inoculated with a normal mesophilic starter and, after a period for ripening, a solution of standard rennet is added. Some 45 minutes later, the curd is cut into cubes and allowed to stand undisturbed for 15 minutes. The curd is then stirred and, over the next 30 minutes, the temperature of the vat is raised to 38 °C; the contents are held at this temperature for a further 30 minutes, with intermittent stirring. The whey is then removed, and the curds are allowed to lie on the bottom of the vat for a further 60 minutes to attain the correct acidity – approximately pH 5.3. Occasional stirring prevents the curd from matting together, as does the addition of salt at a rate of around 2.5%.

Figure 4.9 Majorero, produced on Fuerteventura Island from raw goat's milk. (Courtesy of the Ministerio de Agricultura Pesca y Alimentacion, Spain.)

Figure 4.10 Monterey, a traditional semi-hard cheese that originated in Monterey County, USA; first made by a Mr Jack, hence the popular name of Monterey Jack. (Courtesy of the Wisconsin Milk Marketing Board.)

Individual cheesecloths steeped in hot salt water, or nowadays cloth-lined moulds, are then filled with the curd and lightly pressed overnight. Next morning, the cheeses are removed from the cloths and allowed to air-dry and form a delicate rind, prior to being coated in wax or packed in film. Maturation times vary considerably according to the temperature of storage, but between one and two months may be adequate for the fresh cheese, and up to six months for a skimmed-milk cheese desired for grating. The comparatively high moisture content of Monterey (up to 44%) that results from the light pressing gives a semi-hard texture with a large number of fissures.

Wensleydale

Wensleydale was first made in the Yorkshire dales by monks who came to England with the Normans, and who were then rewarded with the freedom to colonize large areas of land and build monasteries (Davis, 1976). The terrain proved ideal for sheep-rearing, and cheese production became established rapidly. As the monasteries lost their power and cows replaced sheep as the major source of milk, so Wensleydale lost its identity. However, around 100 years ago the name was revived, and it is now officially recognized in the UK Cheese Regulations.

The equipment is essentially the same as that employed for most English cheeses. The milk is pasteurized at 68 °C for 15 seconds before being poured into the vat at 30 °C. A typical mesophilic starter culture is added (0.5–1.0 v/v). Once the acidity has begun to develop, rennet is added at the rate of around 30 ml/100 litres of milk; this level of addition ensures coagulation in 50–60 minutes (Scott, 1986). After cutting, the curd is lightly scalded at 32–34 °C and gently stirred for a further 20–30 minutes. This gentle heat treatment restricts the contraction of the curd and so allows it to retain more moisture than a curd held at a higher temperature. As a result, the subsequent drainage of the whey tends to be quite slow. However, as the texture of the finished cheese should be distinctly crumbly, the coalescing curd is broken at regular intervals into small pieces, a stage that both modifies the structure and assists any free whey to escape.

Once the acidity of the whey has reached around 0.5–0.55% lactic acid, the curd is milled and salted (1.5–2.0%) prior to transfer to moulds capable of holding between 1.5 and 5.0 kg of curd – depending upon the preference of the manufacturer. The filled mould is then allowed to stand overnight at ambient temperature for the curd to settle naturally. On the following day, a light pressing gives the cheese its final coherent structure (Figure 4.11), and it is then ready for maturation at 10 °C for about one month. The mild, acidic taste of Wensleydale is extremely pleasant, but the high moisture content (up to 46% maximum) gives a product that not only matures quickly but also can deteriorate with the same ease; the minimum FDM is 48%. Consequently, it is not a product that is manufactured as widely as might be desirable, and probably few consumers are acquainted with this most attractive of cheeses.

Figure 4.11 Wensleydale, a mild, fast-maturing English regional cheese. (Courtesy of The Cheeses of England and Wales Information Bureau, UK.)

Figure 4.12 Kashkaval, one of the best-known cheeses from Bulgaria; the 'Balkan' trade-mark confirms that it is of top quality and manufactured from full-cream sheep's milk.

Italian-style cheeses

There are a number of cheeses of Italian origin, such as Mozzarella or Provolone, which are placed in a category known as 'pasta filata' cheeses. When young, the nature of the cheeses is semi-soft/semi-hard, whilst the texture is smooth and pliable. These qualities arise mainly from the processing stages, which involve acidification of the curd to a pH of around 5.0, followed by mechanical stretching of blocks of curd heated to at least 55 °C in water or brine (Prato, 1993). This latter action causes the casein to become slightly fibrous so that, on cooling, the resultant cheese takes on the malleable/sliceable texture that is so characteristic of this genre. If taken to extremes, the fibres can be made to take on a firm, distinct character, and the plaited cheeses of the Sudan or the Lebanon (Figure 4.21) provide a classic example of this extreme situation.

Kashkaval

Kashkaval is a typical plastic-curd cheese from the Balkans. While it is traditionally made with sheep's milk (Figure 4.12), cow's milk is often employed to ensure that enough product is available to meet market demand (Figure 4.13). A wide variety of systems are used for production but the essential steps are:

- The addition of a starter and/or rennet preparation.
- The formation of a curd which is then drained and ripened, often tied in cheesecloth, until the acidity has risen and a firm structure has been attained.
- The transfer of the blocks of curd into some form of container for heating until it becomes extremely elastic.
- The moulding of the curd in metal or wooden forms – typically round moulds for Kashkaval of 120–200 cm in diameter and 70–80 cm in depth.

On cooling, the cheeses will be salted in brine for a few hours before surface-drying and packing in film. Dry salting and storage in brine are

Figure 4.13 'Vitosha' brand of Kashkaval, manufactured from full-cream cow's milk.

other options, and the choice depends on the scale of the operation and the intended market. Consumption at one month from production is quite feasible, but maturation for 2–3 months at ambient temperature is more likely.

A typical analysis of Kashkaval might be:

Protein	24–28%
Fat	14–31%
Moisture	36–50%
Salt	4.0%

The wide variations are explained by the fact that some producers remove a portion of the fat from the sheep's milk before cheesemaking.

Around the Mediterranean, this technique of curd manipulation has been exploited to produce a number of related cheeses, and two forms in particular clearly demonstrate the curd elasticity that can be achieved. In the production of a flat cheese that is widely used throughout the

(a)

(b)

Figure 4.14 Production of plastic-curd cheeses. (a) Pieces of curd that have been removed from the cheesecloth and sliced in readiness for the heating stage. (b) On gentle warming in any convenient vessel, the curd becomes sufficiently 'plastic' to be drawn out into long strands. (Courtesy of Dr Imad Toufeili, American University of Beirut, Lebanon.)

Figure 4.15 (opposite) Production of a flat cheese. (a) A lump of curd is placed upon the heated metal plate. (b) The block of curd is then kneaded and stretched. (c) The final stage with the thin sheet of cheese covering the entire surface of the plate. (Courtesy of Dr Imad Toufeili, American University of Beirut, Lebanon.)

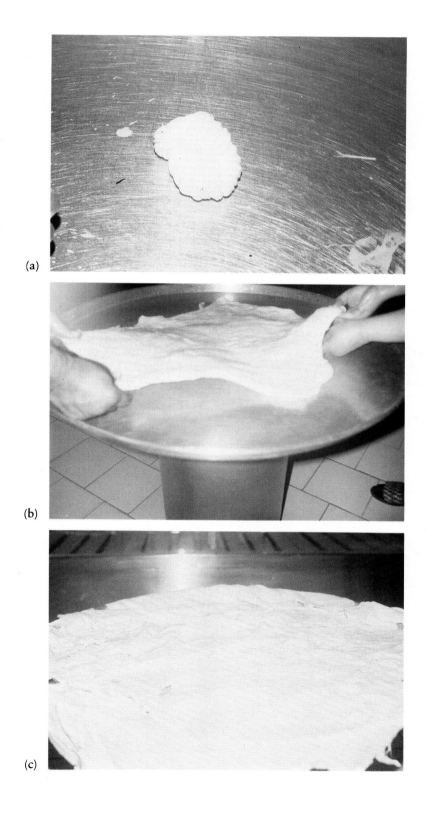

(a)

(b)

(c)

Figure 4.16 Curd manipulation. (a) The 'plastic' curd being drawn and separated to give two strands on each side of the 'loop'. (b) The curd being separated further to give four strands on each side of the 'loop'. (c) The finished cheese, composed of numerous strands. (Courtesy of Dr Imad Toufeili, American University of Beirut, Lebanon.)

coarse plait shown in Figure 4.17, but in either case it is the elasticity of the heated, acid curd that both enables the cheesemaker to perform his/her 'art' and provides the finished cheese with its firm but pliable texture.

Middle East for the preparation of desserts, a lump of curd (Figure 4.15(a)) is placed on a large warm metal plate and gradually stretched (Figure 4.15(b)) until it has become a thin sheet covering the entire plate (Figure 4.15(c)).

An alternative approach is to warm the curd as shown in Figure 4.14(b), and then draw it out into two strands (Figure 4.16(a)), then four strands (Figure 4.16(b)) until the entire block of curd has been reduced to thin 'strings' that are 'knotted' at one end to form the final cheese (Figure 4.16(c)). A less demanding form is the

Figure 4.17 The same plastic curd can be moulded easily into a ribbon and formed into a coarse plait. (Courtesy of Dr Imad Toufeili, American University of Beirut, Lebanon.)

Kasseri

Kasseri is a variant of Kashkaval that is widely popular in Greece, and it is estimated that around 8500 tonnes are manufactured annually from sheep's milk or a mixture of sheep's and goat's milk (Anifantakis, 1991).

Sheep's milk, often with up to 25% of the fat removed, is heated to 32 °C and sufficient rennet is added for coagulation to take place in 45 minutes. Once the gel has formed, it is cut into cubes with sides of around 1.0 cm, and allowed to settle to the bottom of the vat. The whey is removed and the curds are encouraged to mat together with light manual pressure. Blocks of curd are then removed to a draining table, covered with cheesecloth and weighted boards and left to stand for up to 24 hours for the desired level of acidity to develop. After being cut into thin slices, the ripened curd is dipped into water at 75–80 °C and kneaded into a smooth, elastic mass. A portion sufficient to form one cheese is removed and eased into a mould to give a finished cheese of the required dimensions (Figure 4.18). Next morning, dry salting with medium grain salt begins and is continued over a number of days until the number of applications may have reached 10–14; in large factories, one application during the kneading stage has replaced this rather labour-intensive approach. The cheeses are often piled in groups of 6–7 cheeses during the short ripening stage.

The composition of the finished cheese will be:

Protein	26%
Fat	25–29%
Fat-in-dry-matter	32–52%
Moisture	35–42%
Salt	2.6–3.2%

Figure 4.18 Kasseri, a semi-hard cheese made from sheep's or goat's milk in the 'pasta filata' style. (Courtesy of Professor E.M. Anifantakis.)

Mozzarella

According to Prato (1993), Mozzarella was first produced in the area of Naples in medieval times. Buffalo milk was the raw material but, because the conditions of production were usually primitive in the extreme, it would have been quite acidic by the time it reached the cheese vats. Consequently, the finished cheese acquired the malleable qualities associated with Mozzarella, a name derived from the action of the cheesemaker who would 'slice' lumps of the curd into fist-sized pieces for hand-moulding into balls of 100–300 g ready for retail sale.

Traditional Mozzarella is still to be found in Italy, and the Mozzarella di bufalo shown in Figure 4.19 is much prized for its soft texture and delicate flavour. The thin, soft rind is also characteristic of a good quality product and, although edible, it should peel easily off the main body of the cheese. However, factory-made Mozzarella is much more likely to be in a block form (Figure 4.20) wrapped in polyethylene, and to be of the so-called 'pizza' variety. This pizza or low-moisture form is processed to give a product with

Figure 4.20 Low-moisture Mozzarella, derived from the Italian variety and increasingly popular in the USA. (Courtesy of the Wisconsin Milk Marketing Board.)

a compact, firm texture and precise properties with respect to melting; this latter feature, which is vital for a pizza topping, is also affected by salt content, so that pizza Mozzarella has a defined salt content of 1.5–1.7% compared with <1.0% for Mozzarella di bufalo.

The standard method for manufacturing Mozzarella involves first standardizing the milk with respect to fat content – perhaps 3.6% for fresh Mozzarella and 1.8% for the pizza variety – followed by pasteurization. Inoculation with *Str. thermophilus* (1–2% v/v at >30 °C) is employed to raise the acidity of the milk. For pizza Mozzarella, strains of *Lac. bulgaricus* are used alongside the streptococci. The inclusion of this latter species is valuable for achieving a rapid increase in the acidity of the milk (up to 0.4% for the pizza cheese at the time of coagulation), but serves an additional function in that the *Lactobacillus* shows marked proteolytic activity; the partial modification of the casein fraction can be used to adjust the melting and other properties of the finished cheese.

Once the milk has been acidified to the correct degree, a low level of rennet can achieve coagula-

Figure 4.19 Mozzarella di Bufalo, a traditional variety made in Campania and Latium from buffalo milk. (Courtesy of the Ministero dell'Agricoltura e delle Foreste, Rome.)

tion in around 30 minutes. The curd is then cut into large pieces (sides of 4–5 cm) and after a rest period the curd is cut again so that the size of the cubes is roughly halved. In some factories, the curd is allowed to stand for a while under the whey, but otherwise it is placed on draining tables for 3–4 hours (22–25 °C) until the pH has reached around 5.0. At this point, the curd is transferred to hot water at 70–80 °C for manual or, more usually, mechanical stretching and kneading. The extent of the physical stretching and folding is determined by the type of cheese, so that while fresh Mozzarella may be exposed to a minimal treatment, the pizza type is treated much more severely. This contrast is reflected in the nature of the finished cheeses, for while the fresh variety is comparatively soft, the pizza type has to be firm enough for easy slicing or grating. This same desired difference influences the salting process, in that while pizza cheese is often salted during the stretching and forming stage, the fresh cheese is always salted later.

Once the correct texture has been achieved, the cheese will be machine or hand moulded. At this point, the temperature of the curd will have fallen to <60 °C. It must not be allowed to remain at this temperature and should be cooled as quickly as possible with chilled water. Cold brine can be used to the same effect with fresh Mozzarella, so achieving cooling and salting in one stage. Packaging of the retail portions follows – often in 'preserving fluid' under vacuum, but much pizza cheese is distributed as shrink-wrapped, catering-size blocks of several kilograms. The 'preserving fluid' is usually the water that was used to heat the cheese during the stretching operation, and the residue of starter bacteria is alleged to provide protection against spoilage and/or pathogenic bacteria.

The finished cheese has the following composition:

Protein	18–22%
Fat	22–27%
Fat-in-dry-matter	58–59%
Moisture	54–62%

The variations are accounted for by the type of cheese and/or the source of the milk (i.e. buffalo or cow). A further source of variation can be the process itself, for a recent modification of the system involves souring of the milk with citric acid rather than with starter cultures. The aim is to standardize the process and avoid the vagaries introduced by unpredictable starter activity, and certainly for pizza cheese the system has become popular in America. Whether or not the organoleptic quality of a fresh cheese can match that of a traditional Mozzarella is a moot point, but it is recorded by Prato (1993) that this chemical souring process is to be found even in Italy. One further contrast is the absence of 'preserving fluid', so that a special solution – including preservatives – has had to be formulated to ensure the absence of any risk to public health, e.g. survival of *Listeria* (Papageorgiou and Marth, 1989).

Ostiepok

This cheese has a long history of production in various regions of Czechoslovakia (as it then was) as it combines both simplicity of manufacture and stability at ambient temperature. Sheep's milk is poured into any convenient vat, and then renneted at 32–34 °C. Further treatment is similar to that accorded to the milk for the production of Hrudka (see Bryndza cheese), except that the collected curd is broken, kneaded and forced into convenient draining dishes rather than being hung in cloths. Once the whey has drained out, the moulded cheese is dipped into water at 60 °C until it takes on the desired 'elastic' texture. At this point, the cheese is removed from the dish and, while still warm, manipulated into the final shape or eased into ornamental moulds of any chosen form (Figure 4.21).

After cooling, the cheeses are placed in 15–20% brine for 24 hours or so, before being hung in a smoking room for at least one week. The final cheese has, according to Davis (1976), a moisture content of 35–40%, an FDM value of 42–47% and a salt content of 3.0–3.5%. The comparatively heavy deposition of smoke gives the cheeses its distinctive coloration and flavour,

Figure 4.21 Ostiepok, originally produced in mountain regions from sheep's milk but now manufactured industrially from cow's milk. (Courtesy of the Dairy Research Institute, Zilina.)

Figure 4.22 Smoked roll-cheeses have always been popular in Central Europe, and the flavour of chopped ham blends well with that of the natural cheese. (Courtesy of CMA (UK), London.)

and while the flavour of traditional Ostiepok may be too strong for many tastes, the more mild variant manufactured from cow's milk is not unlike many smoked cheeses found throughout central Europe. A typical modern form is shown in Figure 4.22, and with industrial varieties of this type a mild smoking is employed purely to enhance the flavour rather than for any preservative function.

Provolone

Although the growing popularity of pizzas has made most people aware of Mozzarella, Provolone is a variety of equal stature amongst Italians. Hailing from southern Italy, it is now manufactured in many countries, and the characteristic pear-shaped cheese (Figure 4.23) is the most widely marketed form. The basic cheese is sold in two forms: Provolone dulce (a mild flavoured variant) and Provolone piccanti. The latter is made with a crude type of rennet that includes lipases as well as proteases in its formulation, so that the resultant cheese acquires a much stronger taste.

It is manufactured from full-cream cow's milk. After pasteurization, the milk is cooled to around 36–37 °C, ready for inoculation with a mixture of thermotolerant streptococci and, often, *Lac. delbreuckii* subsp. *bulgaricus*. As the acid develops, a comparatively high level of rennet is added – 25 ml/100 litres of milk. At the elevated temperature, clotting occurs in about 15 minutes. The coagulum is then cut into fine cubes and, as the whey drains off, the curd is spread out along the floor of the vat or on to a draining table. After several hours at 30–35 °C, the curd has coalesced into a solid sheet some 7–10 cm thick. It is then cut into pieces about 20 cm wide and 60 cm or so in length. The timing of this slicing stage depends, in the main, on the ambient temperature and/or the activity of the starter culture, because a well developed acidity is essential for the final stages of processing.

In some factories, the resting curd is tested at intervals by placing a small piece into hot water at 80 °C, and then manipulating it to test whether it will stretch without breaking. In any event, the slices of curd will be allowed to stand for a further period prior to being immersed in water at 55 °C.

Once the curd has reached the desired temperature, it is kneaded either mechanically or manually until it achieves the elastic texture that is so characteristic of 'pasta filata' cheeses. It may take 15–20 minutes for the curd to achieve the correct texture and at this point the mass is removed from the water and cut into pieces suitable for final moulding as individual cheeses. The sizes of the retail cheese can vary from factory to factory (4–5 kg might be regarded as typical) but the rind should always be smooth and shiny in appearance, and with a colour that ranges from pale yellow to golden brown. The interior of the cheese should be white or pale yellow; the flavour, at around two months, should be mild and creamy, but with a delicate hint of 'cheese'. A

Figure 4.23 Provolone, a soft to medium-hard cheese produced widely in Italy from cow's milk. (Courtesy of the Ministero dell'Agricoltura e delle Foreste, Rome.)

stronger flavour develops at 4–6 months, but maturation beyond this point is undesirable due to water loss and a hardening of the texture; old Provolone is often used as a grating cheese.

A typical analysis of Provolone might be:

Moisture	37–43% (max. 45%)
Protein	28%
Fat	25–33%
Fat-in-dry-matter	47% (min. 44%)
Salt	2–4%

White-brined cheeses

Along the shores of the Mediterranean, cheese has been produced for thousands of years and a number of features of the region have tended to dominate the systems of production. In particular, the following aspects were of especial relevance:

- The use of sheep's milk, which led to the expectation that the cheese would be white in appearance.

- The high ambient temperatures, which favoured the growth of natural thermophilic bacteria during the ripening stage.
- The need to preserve the cheese from rapid spoilage by immersion in brine.

Consequently, a whole family of 'white-brined cheeses' evolved which, over the centuries, differentiated into the varieties that are recognized today. Many local cheeses have remained as little more than 'white cheeses stored in brine', but others like Feta or Domiati have acquired international recognition. However, the position is complicated by the fact that many industrialized countries away from the Mediterranean now manufacture 'Feta' for local sale and/or export, and apply the name to products that range from 'soft and spreadable' through to 'hard and sliceable'. Attempts to restrict the application of the name to 'good quality, brined cheese in Greece from sheep's milk or a mixture of sheep's and goat's milk' have met with little success to date, even though many people would accept that a historical link does exist between the country and the cheese.

Domiati

Domiati is the traditional white-brined cheese of Egypt, and is believed to have originated there some time after 332 BC (Abou-Donia, 1991). It is not unlike Feta cheese in many respects, and nowadays is even marketed in the same square metal cans, although small dairies still produce the cheese as the more familiar round – 6 cm or 12 cm × 4 cm (Figure 4.24). It has a mild, salty flavour when fresh, but can become quite acidic after a few months' maturation. As might be expected, the texture becomes firm and slightly 'flaky' after about three months in brine.

The method of manufacture is one of the remarkable features of Domiati. It begins with the standardization of buffalo's or cow's milk – or a mixture of the two – to give milks of 8%, 4% or 2% fat, the choice being governed by the type of cheese desired by the market, i.e. full cream or reduced fat. The standardized milk is

Figure 4.24 Domiati, one of the best-known traditional cheeses of Egypt, is unusual in that the salt is added to the milk prior to manufacture; storage in brine may increase the level of salt in the finished cheese. (Courtesy of Professor S. A. Abou-Donia.)

then divided. Around one-third is heated to 80 °C; the remainder is salted to a level of between 5% and 14%, with the precise rate being determined largely by the season. This early salting of the milk is characteristic of the Domiati process, and it plays a crucial role in preventing the spoilage problems that might otherwise arise from the use of milk of indifferent bacteriological quality. The advent of pasteurization would allow manufacturers to change the system, but Domiati has always been made with salted milk and the dairy industry in Egypt sees no reason to change this time-honoured custom.

After salting, the two portions of milk are recombined to give a temperature of around 38 °C, and the milk is renneted and inoculated with a starter culture of salt-tolerant lactobacilli and/or other genera; the choice of starter seems to vary from factory to factory. Over the next 2–3 hours, the combined action of acid and rennet causes the milk to gel, and the firm coagulum can be ladled out into moulds lined with cheesecloth or some similar material. Where large moulds are employed, the curd may be lightly pressed to expel the whey; otherwise, natural drainage and frequent turning of the moulds is sufficient. After 24 hours the moulds are removed and the cheeses cut into convenient blocks for retail sale as fresh cheese. Alternatively, the blocks can be stored in salted whey for several months to acquire a stronger flavour, or be sealed in square metal cans with brine to occupy the spaces between the blocks.

Some typical compositions of Domiati at four months old are shown in Table 4.1.

Table 4.1 Typical compositions (%) of Domiati

	Cow's milk	Buffalo's milk
Moisture	55.0	55.0
Fat	20.0	25.0
Protein	12.9	11.8
Fat-in-dry-matter	44.4	52.1
Salt	4.9	4.8

Feta cheese

As defined by Anifantakis (1991), Feta is a white cheese, ripened and kept in brine and having a salty, slightly acid taste and crumbly texture. Around 140 000 tonnes of this traditional cheese are produced annually in Greece, but world-wide production employing milk concentrated by ultra-filtration is even more extensive. However, it is the Greek cheese that is usually regarded as the 'bench-mark' product; hence it is the manufacture of this cheese that deserves detailed consideration.

The best milk for making Feta cheese is sheep's milk, although mixtures with goat's milk can give a good quality product as long as the level of goat's milk does not exceed 30%. The use of cow's milk is permitted in Greece to cope with periods of high demand, but it is claimed that the fine qualities of true Feta are somewhat lacking. In the absence of on-farm refrigeration, the sheep's milk is delivered to the dairy once a day during winter and twice a day during summer to

ensure that acid development does not take place prior to pasteurization. On arrival at the dairy, the fat content of the milk will be standardized to around 6.0% – it varies between 6.5% and 8.0% according to the season – and the milk is then pasteurized at 72 °C for 15 seconds or an equivalent treatment. Some Feta is still manufactured with raw milk but, for public health reasons, the practice is becoming less common.

On cooling to 32–34 °C, a starter culture is added. Most dairies employ a normal yoghurt culture at a rate of around 0.5% (v/v) to encourage a fairly rapid and extensive acid production. The type of rennet is another distinctive feature of the process: the addition consists of three parts of standard rennet to one part local rennet. This latter product is made by taking the dried abomosum of a young lamb or goat that has been slaughtered before weaning, and then soaking it in brine (10%) for about 24 hours. The brine, when it is filtered off, contains a high level of chymosin as well as other enzymes, such as lipase, that contribute to the development of flavour in the finished cheese. Calcium chloride is often added at the same time to obviate any possible deficiency of calcium ions, and the milk is then allowed to stand for 20–30 minutes to coagulate. Cutting into cubes with sides of 2–3 cm follows, but there is no use of scalding to accelerate the loss of whey from the curd.

The moulds are circular units of stainless steel or polypropylene, with fine slits cut at regular intervals down the sides and across the base to facilitate whey drainage. The precise dimensions of the moulds tends to vary quite widely. The soft curd is ladled into them in stages, because if the entire mould is filled all at once, whey tends to remain trapped and cause poor flavour development or spoilage problems. After 2–3 hours, the lidded moulds are inverted for a further period to continue the draining process. The exact timing is dependent on the temperature – around 15 °C is preferred. Once the cheese has become firm enough to be removed from the mould, it is cut into blocks and dry salted using coarse-grained salt. The choice of this material is regarded as essential, in that while the large granules dissolve slowly and draw whey from the curd in an even manner, fine salt would tend to dry and harden only the surface layer of the cheese so that draining would then be slow and unsatisfactory. At one time, dry salting was continued over a number of days, after which the blocks were left on the salting table – with occasional turning – for a layer of yeasts and bacteria to develop over the surface. This microflora was credited with the ability to enrich the flavour of the Feta, but nowadays the pressures of increased production have meant that the salting stage has been modified or, in some cases, replaced by a direct transfer from mould to brining tank. However, a common compromise involves dry salting and

then immediate transfer to wooden barrels for ease of storage (Figure 4.25). Occasional turning and resalting of the blocks can still be accomplished, and after about a week the barrels can be filled with brine (7–8%) for longer term storage.

During the dry salting and/or initial brining, the temperature is maintained (as far as possible) at 14–16 °C, but once the pH of the cheese has fallen to around 4.5, the barrels are moved to cold stores at 3–4 °C. At least two months' maturation should be allowed before retail sale is contemplated, but the stability of the brined cheese allows for long-term storage as well. This latter property is exploited with the packing of Feta in metal cans – once sealed, the containers can be handled with ease.

The typical compositions of Greek Feta and the closely related Telemes cheese are shown in Table 4.2. All figures relate to products manufactured from sheep's milk.

The main differences between Feta and Telemes are that:

- the moulded curd for Telemes is subject to pressure to expel the whey, while the whey drains from Feta curd by gravity and the action of the coarse salt; and
- Telemes is salted by immersion in strong brine after removal from the moulds and cutting into blocks of around 1 kg.

However, although Greek legislation does include individual standards for both cheeses, it may be that the average visitor to Greece would not be immediately aware of the contrast.

Figure 4.25 Feta, the traditional white-brined cheese of Greece, originally made only from sheep's milk and traded in wooden barrels. (Courtesy of Professor E. M. Anifantakis.)

Table 4.2 Typical compositions (%) of Feta and Telemes

	Feta	Telemes
Moisture	52.9	52.2
Fat	26.2	n/a
Protein	16.7	17.6
Salt	3.0	3.5
pH	4.4	4.7

After Anifantakis (1991).

White-brined cheese

As already mentioned, Feta is now manufactured world-wide from cow's milk – either direct or from milk that has been concentrated by membrane processing (Grandison and Glover, 1993), and even from recombined milk in some parts of the world.

Vast tonnages of Feta-style cheeses are also manufactured. While some forms are distinct varieties (e.g. the Lightvan cheese from Iran), many are simply 'white cheeses that are stored in brine'. However, one type that has achieved a degree of recognition is the White-brined cheese from Bulgaria (Figure 4.26). This traditional cheese is sold widely throughout the Balkans and the Middle East, and the range of packages (from parchment paper to sealed tins containing several kilograms of cheese) confirms that both local and export uses are envisaged.

Figure 4.26 The dominant cheese in Bulgaria, referred to simply as White-brined cheese, is similar in style to Feta. Traditionally it was all manufactured from sheep's milk, but some is now produced from cow's milk. Distribution in cans of brine or salted whey is common.

5 Cheeses with visible 'eyes' in the structure

Although basically hard or semi-hard in texture, there are many varieties of cheese that are, on slicing, quite distinctive in appearance in that the cut surface reveals the presence of many round openings – often called 'eyes' – that break up the uniform structure of the curd. The size and number of these openings, and indeed their very presence or absence, depends upon two essential factors:

- the presence of specific genera of bacteria in the original starter culture that liberate carbon dioxide (CO_2) during the maturation of the finished cheese; and
- the physical properties of the curd being such that the CO_2 is trapped in the curd and forms a visible cavity.

The latter property is determined by the actual cheesemaking procedure itself, but the size and number of the eyes is determined, in the main, by the types of bacteria present. Thus, if the openings are generated by heterofermentative species of *Lactococcus* or *Leuconostoc*, such as *Lact. lactis* biovar *diacetylactis* or *Leu. mesenteroides* subsp. *cremoris*, then their metabolic activity tends to be both early in the maturation stage and somewhat limited in extent. Consequently, cheeses that employ these starter cultures, such as Edam or Gouda, may have only a few rather small eyes.

Swiss-style cheeses, by contrast, have large, conspicuous openings, and Emmental is a classic example of the genre. Some Danish varieties have equally distinctive cavities in the curd, but the association between dimensions of the eyes and the type of starter culture is not necessarily conclusive. One Italian cheese, Fontina, is a case in point, for the small openings in the curd of this cheese are a reflection of the process rather than any lack of bacterial activity *per se*.

The organisms responsible for gas formation in the Swiss-style cheeses are the so-called propionic acid bacteria, and strains of *Propionibacterium freudenreichii* are the most important. This species, in contrast with the lactococci for example, grows only slowly in milk, but is able to continue its activity well into the maturation stage. This ability is linked, in part, with its capacity to metabolize lactic acid to propionic and acetic acids, and CO_2, so that the bacteria are able to find a readily available substrate in the cheese. Their slow growth is alleviated to some extent by the synergistic activities of the other starter bacteria associated with Swiss cheese, namel *Lac. bulgaricus* and *Lac. helveticus*, but even so the performance of the propionibacteria can, under industrial conditions, be somewhat erratic. It is for this reason that the production of a wide range of cheeses nowadays involves the use of heterofermentative lactic acid bacteria, and it may be that the character of some common varieties will alter as a result.

The other crucial property of these cheeses is the flexible nature of the curd, and this feature is dependent on the use of fast acid-producing cultures that can withstand the high scald temperatures needed to modify the curd. Salting in brine is essential also to avoid inhibition of the salt-sensitive propionibacteria, so that the salt does not (*vis-à-vis* a hard cheese, for example) play a dominant role in texture formation.

Appenzeller

This flat, round cheese – approximately 20–30 cm in diameter and 10–15 cm in height – is named after the Canton of Appenzell in Switzerland. The rind is firm, and often acquires a yellow–brown appearance as the result of bacterial activity (Chapter 7), whilst the interior is white to pale yellow. The texture is soft for this category, but the cheese is still sliceable, and firm enough to retain sufficient gas to produce a few well-defined eyes (Figure 5.1).

A mixture of cow's milk from the morning and evening milkings provides the basic raw material and, for some brands, it may be partially skimmed before pouring into the vat. Heat treatment is usually quoted as 'none' (Scott, 1986), so that the next essential is to regulate the temperature to 30 °C prior to the addition of the starter culture. A standard cheese culture is used to provide the acidity, with propionibacteria from the environment or the culture providing the source of gas for eye formation. Sufficient rennet to give a firm gel in around 30 minutes is added – perhaps 25 ml/100 litres of milk – and the coagulum is then cut and slowly stirred. Scalding takes place at 43–45 °C for full-fat curds or 7–8 °C lower for curds derived from skimmed milk. Once the curd has acquired a firm texture, the whey is drained off. The curd is then transferred to moulds of the chosen dimensions, and pressure is applied for about one hour or until the curd has taken on the shape desired for the finished cheese.

Standing overnight at ambient temperature helps to consolidate the curd, and the cheeses can then be placed in a bath of brine (20% salt) for 2–3 days. Holding in the cheese store for about 10 days allows the cheeses to dry and form a soft rind, while the growth of bacteria and yeasts on the surface may be instrumental in enriching the developing flavour. Once the cheeses are dry and the rind is firm, the distinctive stage in the processing is introduced: the cheeses are immersed in cider or spiced white wine, or coated with a mixture of salt and spices suspended in white wine. The length of exposure to, and precise choice of, seasoning materials is at the discretion of the cheesemaker, but the end result is a cheese with a quite distinctive flavour. The full-fat cheese (50% FDM) tends to be more mild in flavour than the low-fat variety, and the eyes tend to be smaller and more numerous.

Figure 5.1 Appenzeller, a flat, cylindrical cheese which is so named because one of the ripening stages involves immersion in cider or spiced white wine. A few round holes should be present. (Courtesy of 'Cheeses from Switzerland'.)

Beaufort

According to early records, a form of Gruyère-type cheese was made in France as early as 1267 (Davis, 1976). Today, over 100 000 tonnes of Emmental, Comte (Gruyère de Comte) and Beaufort cheese or 'Mountain Gruyère' are produced annually. French Emmental is very similar to its more famous counterpart from Switzerland, but both Comte and, more especially, Beaufort have characteristics that makes them distinctive to the connoisseur.

Of the two cheeses, the production of Comte has become the more commercial, and large dairies now control most of the output. Full-cream milk is always used, and the starter culture will include both thermophilic lactobacilli and propionibacteria. A high scalding temperature – around 56 °C – is used to ensure that the curd acquires the flexible texture essential to allow proper eye formation. The finished cheese, which may be up to 45 kg in weight, is rubbed with brine and/or dry salt after pressing, and then subjected to maturation for at least 4 months at 18–20 °C. During this time, a few openings of 1–1.5 cm develop, and this restricted activity is, in part, a consequence of the losses of propionibacteria that occur during scalding. The finished cheese is hard, in comparison with similar cheeses, with a moisture content below 38% and a minimum FDM of 45%.

In some respects, Beaufort could be classified just as easily as a semi-hard cheese, because the frequent lack of visible eyes is a direct contrast with many other cheeses in this group. However, the presence or absence of openings is more a reflection of the lack of process control than anything else, hence it is probably fair to include the variety alongside its namesake.

It is made only in the alpine regions of Savoie during the summer months when the cows can graze the high pastures, and mountain huts with wood-fired stoves provide the bases in which the cheeses are made. The high fat content of the cheese – often in excess of 50% FDM – follows from the absence of any adjustment of milk composition, but the advantage is that the finished cheese has a texture that is more supple and waxy than normal Gruyère. Similarly, gas formation during maturation will depend on the fortuitous presence of heterofermentative lactococci and/or leuconostocs rather than propionibacteria, and yet the specific origin of the milk and the traditional approach to cheesemaking combine to give a totally unique product. Once pressed, the cheese, which may be up to 50 kg in weight, is rubbed with brine and dry salt, and the characteristic colour of the coat (Figure 5.2) suggests that the superficial activity of yeasts and bacteria contributes to the distinctive flavour and aroma of the cheese.

At the end of the season, the cheeses are brought down to a central store in Beaufort for final maturation. Much of the limited production – only a few hundred tonnes/year – is sold locally but some is marketed further afield.

Figure 5.2 The full name for Beaufort cheese – Le Gruyère de Haute Montagne Beaufort – indicates quite clearly the nature of this cheese, except that the well-defined 'eyes' are lacking. (Courtesy of CIDIL, Paris.)

Colonia and Holanda

Holanda or Pategras is the most common semi-hard cheese in Argentina and, along with Colonia, provides a good example of the migration of an essentially European cheese to a new environment. Holanda is round in shape (Figure 5.3) and has a very elastic texture and pleasant, sweet flavour. Full-cream cow's milk provides the raw material and, although it may be pasteurized to eliminate pathogenic organisms, the low temperature batch system is the preferred option. The reason for this selection is that eye formation

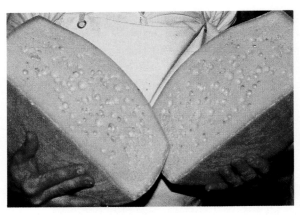

Figure 5.4 Colonia, another Dutch-style cheese but larger and longer maturing than Holanda. (Courtesy of Dr Carlos Zalazar, Instituto de Lactologia Industrial, Argentina.)

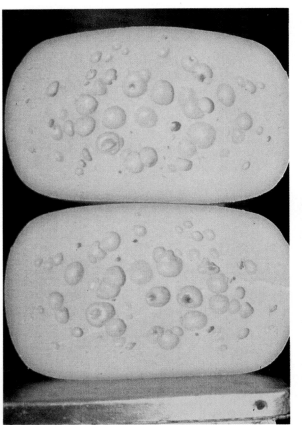

Figure 5.3 Holanda or Pategras cheese, widely consumed in Argentina, has much in common with the Dutch varieties. (Courtesy of Dr Carlos Zalazar, Instituto de Lactologia Industrial, Argentina.)

is dependent upon the presence of leuconostocs and other heterofermentative bacteria in the raw milk, and hence any heat-treatment must be of limited intensity. This natural microflora, along with the standard rennet, provides the basic coagulum, while the continued production of gas during ripening gives rise to the round eyes – about 1.5 cm in diameter – that are evenly distributed across the finished cheese. The maturation time is around one month, and a typical cheese weighs 4 kg.

This latter point provides a contrast with Colonia, which is more likely to be in the region of 7 kg (Figure 5.4). It is broadly similar to Holanda, but tends to be regarded as superior in quality because the raw milk is often delivered only from specifically selected farms. It is matured for around two months to ensure that its organoleptic qualities are fully developed.

Danbo

Danbo is a popular cheese in Denmark, and it displays the pale interior typical of many such cheeses, along with a few, very distinct eyes (Figure 5.5). The formation of these openings is encouraged by the use of a modest temperature for scalding, in that it is sufficient to modify the texture of the curd and give a finished cheese with a supple and sliceable body, but not so high as to eradicate the heterofermentative organisms essential for eye formation. A further process stage that encourages microbial development is the practice of pressing the scalded curd beneath the warm whey at the bottom of the vat, an environment that is conducive to bacterial activity. It is relevant also that this same procedure allows the curd particles to coalesce without any entrapment of air, so that the cut surfaces of the finished cheese possess a smooth, even appearance totally devoid of any irregular mechanical fractures.

As with most cheeses of this type, immersion in tanks of brine is sufficient to impart a salt level of around 1.0% to the curd, and the cheeses are then stored at 10–20 °C for the rind to dry and harden. Occasional wiping with a brine-soaked cloth and turning encourages the advance of the maturation process and, after a few weeks, the cheeses are moved to a low temperature store (8–12 °C) for a further 4–6 weeks. After one final clean, the cheese is stamped to show the variety and the FDM value – 45% and 30% are the recognized levels for Danbo, prior to coating with paraffin wax. This wax not only helps to maintain the clean, attractive appearance of the cheese by preventing the growth of fungi or bacteria on the rind, but also avoids moisture loss whilst the cheese is being transported to the retail outlets.

The maximum moisture content of the retail product is set at 47% for the high-fat variety and 53% for the medium-fat cheese. The mild, slightly aromatic flavour of Danbo may be enhanced further by the addition of caraway seeds, and incorporation during manufacture ensures an even distribution throughout the block. Although small retail blocks are available, the traditional cheese weighs around 6 kg.

Figure 5.5 Danbo, a rectangular cheese coated with yellow or sometimes red cheese wax. The body is firm and sliceable but more supple than many cheeses. A few large, round holes are clearly visible. (Courtesy of the Danish Dairy Board, Aarhus, Denmark.)

Edam

Edam is one of the most famous cheeses of Dutch origin, and its distinctive spherical shape and, at least outside Holland, bright red, wax coating (Figure 5.6) make the variety instantly recognizable.

Milk of good microbiological quality is desired for the production of Edam and, after standardizing the fat content to 2.5%, the milk is pasteurized. Calcium chloride may be added in some regions to assist in the coagulation stage, and because Edam is a low-acid cheese, sodium or potassium nitrate may be used to suppress the growth of coliforms and/or the development of any spores of clostridia that may be present. If spores of *Clostridium* spp. germinate and grow during the maturation stage, excessive gas production ('late blowing' of the cheese) and the formation of off-flavours can cause severe losses of product; coliforms can prove a source of

Figure 5.6 The small, spherical Edam (4 kg) has a distinctive coating of red cheese wax (at least for export). The pale yellow cheese is waxy and pliable in body, with a mild flavour. (Courtesy of the Dutch Dairy Bureau, Leatherhead, Surrey, UK.)

unpleasant taints as well. A mesophilic starter culture is added to the milk at around 30 °C, together with some 30 ml of standard rennet/100 litres of milk.

The coagulum should be firm enough for cutting around 30 minutes later and, after a further period of gentle stirring, about 50% of the whey is removed. Warm water at 55–60 °C is then slowly added to the vat, so that the temperature of the curd/whey mixture rises to 37 °C or thereabouts. This 'washing' of the curd has the effect of removing some of the lactose, so reducing the tendency of the cheese to develop acidity during maturation; Edam is essentially a sweet curd cheese with a pH in the region of 5.4. Once the curd has become firm, most of the dilute whey is removed and the partially submerged curd is pressed mechanically so that it mats together with no visible air spaces.

Although block Edam has found favour with some factories, the classic moulds for the cheese are round plastic or metal spheres manufactured in two halves, each around 150 cm in diameter and 150 cm deep. After the moulds have been filled with warm curd (30 °C), the cheeses are exposed to a mild pressure for 2–3 hours in order to give a firm structure to the product. The moulds are then removed, and the cheeses are dipped into hot whey (around 55 °C) to assist in further 'plasticizing' the body of the cheese. Trimming any irregularities from the surface gives the cheese a neat, round appearance, and final pressing overnight expels any residual whey and compacts the curd into a firm, even mass. Immersion in near-saturated brine for 2–3 days is sufficient to give a salt content of 2.5% in the finished cheeses, which are then stored at 14–15 °C (RH of 85–90%) for the rind to dry off and harden. Often the cheeses are piled on to one another to give the upper and lower surfaces a slightly flattened form, and occasional washing/cleaning may be necessary to avoid any risk of microbial activity on the surfaces.

Full maturation should be achieved in 6 weeks at 14–15 °C, but a mature variant (Figure 5.7) with an in-store hold of 6 months is available on the market. The normal cheese has a compact

Figure 5.7 Apart from the traditional Edam, there are herb, cumin and pepper varieties, as well as a mature Edam with a distinctive coat of black wax. (Courtesy of the Dutch Dairy Bureau, Leatherhead, Surrey, UK.)

and 'rubbery' texture, and the small eyes are usually few in number. The flavour is mild, almost to the point of sweet, and herbs, such as cumin or pepper, may be added to provide the consumer with a number of taste options which may, in some cases, be identified by the colour of the wax applied to the surface-dried cheeses. Typical analyses of the cheeses are shown in Table 5.1.

While Edam accounts for some 25% of the cheese produced in Holland, this figure does not take into account those varieties that are closely modelled on the original. Mimolette, for example, is clearly derived from Edam, and only the distinctive orange colour (Figure 5.8) marks it as truly different. It is notable also that Edam-style cheeses enjoy widespread popularity with cheese-makers elsewhere in Europe, and a casual observer might well be confused as to the origin of the Mimolette from northern France (Figure 5.9).

Table 5.1 Typical composition (%) of Edam

	Standard (6 weeks)	Mature (6 months)
Protein	28	31
Fat	24	26
Fat-in-dry-matter	42 (min. 40)	41
Moisture	43 (max. 45)	37
Salt	2.5	2.8

Figure 5.8 Mimolette, known locally as 'Commissiekaas', a spherical cheese with the taste and texture of Edam. Carotene provides the distinctive orange colour. (Courtesy of the Dutch Dairy Bureau, Leatherhead, Surrey, UK.)

Figure 5.9 Mimolette, an Edam-style cheese produced in northern France; regional variations in appearance and size are quite common. (Courtesy of CIDIL, Paris.)

Elbo

Elbo is one of a family of cheeses from Denmark with characteristics that are broadly similar. Samsoe, in particular, has much in common with Elbo, except that Elbo is always rectangular in shape (Figure 5.10). The weights of the cheeses vary from 2 kg upwards, with 5 kg being perhaps the most popular. Local cheeses may be sold with natural, dry, yellowish rinds, but most are marketed with wax coatings of yellow or red. On cutting, the interior is white or yellowish in appearance, and the surface will display a number of regular openings in the order of 0.5–1.0 cm in diameter. The texture is firm and sliceable and, as with many cheeses of this type, the aroma is mild and slightly aromatic. Minimum FDM contents are set at 40% or 45%, depending upon the precise type or brand of cheese, and the corresponding moisture levels will be 47% or 48%.

Figure 5.10 Elbo, another cheese coated with red cheese wax. The regular, round holes are a conspicuous feature. (Courtesy of the Danish Dairy Board, Aarhus, Denmark.)

Emmental

Emmental is the classic Swiss cheese, and the large wheel-shape, together with the presence of many large eyes (Figure 5.11), has made it famous around the world. It was produced originally on farms in the Emmental Valley, but over its history of 300–400 years, manufacture has become based almost entirely in large factories. Even so, many cheesemakers have retained the traditional small vats (around 900 litres capacity) for ease of manipulation of the curd and control of the process conditions.

The milk, at a standard fat content of 2.8–3.0%, is inoculated with a culture consisting of one part *Str. salivarius* subsp. *thermophilus* to two parts *Lac. helveticus*, along with a culture of propionibacteria. At around 30–32 °C, a moderate level of standard rennet is added at 16 ml/100 litres of milk (Davis, 1976) to give a firm coagulum in 20–30 minutes; calcium chloride may be added to ensure that the gel is sufficiently

Figure 5.11 Emmental (Emmentaler), characterized by the extremely large 'eyes'. Although generally regarded as a Swiss cheese, Germany is now a major producer. (Courtesy of CMA (UK), London.)

tough to withstand subsequent handling. Cutting into pieces the size of rice grains follows, and the curd is stirred intermittently as the temperature of the vat is raised to 45 °C over a period of some 30 minutes. This temperature can be tolerated by the starter bacteria but not the mesophiles that may be present, and even the streptococci and propionibacteria will be stressed by the final scald at around 55 °C. However, as the ultimate heating stage only lasts about 10 minutes, sufficient propionibacteria survive within the curd pieces to bring about the essential eye formation.

Once the particles of curd have become firm, the heating and stirring are stopped and some of the whey is drawn off and replaced by cold water. The curd is then collected into a large square of coarsely woven material and mechanically hoisted out of the vat. The weight of curd extracted in this manner is enough for one cheese (Figure 5.12). The entire mass – including the cloth – is then forced into one large, round mould up to 80 cm diameter and 20–25 cm in height.

Pressure is applied over the next two days, often increasing with time, and the cheeses are turned frequently to ensure that the curd fuses evenly and no whey becomes entrapped. At the end of pressing, the cloths are removed and the cheeses, still in their moulds, are taken to a cool room for an initial period (1–2 days) of dry salting. Immersion in concentrated brine follows and the salt content is controlled so that the cheeses just float at the surface of the tank – a situation that is maintained by turning the cheeses daily and sprinkling the exposed surfaces with salt to compensate for that lost by diffusion into the curd. A temperature of 8–10 °C is regarded as optimum, and it is common to find a control system involving the circulation of the brine via a refrigeration unit.

After 2 days, the cheeses are removed from the brine and placed on wooden boards that make it possible to move the large 'wheels' around without the risk of cracks developing in the rind. A short period of cool storage follows, and the

Figure 5.12 A typical Emmental cheese showing the characteristic wheel-shaped form. (Courtesy of Texel, Epernon.)

cheeses are brushed, dry salted and turned daily to encourage the formation of smooth, dry rind. Once dry, the cheeses are placed on shelves in a store held at around 20 °C, and it is during this phase that the propionibacteria become active. A moderate humidity of 80–83% is maintained in the room, as this level prevents the cheeses from drying out, but at the same time curtails the growth of surface moulds; even so, the cheeses have to be brushed and washed with brine on a regular basis to keep the surfaces clean. Some 2 months later, the large eyes will have developed, along with a mild flavour, and the cheeses are ready for sale. The actual size of the openings is determined to some extent by temperature, and some manufacturers hold the cheeses for 1–2 weeks at 25 °C to ensure that the eyes attain their distinctive character.

Although mature at 2 months, much market cheese is likely to be around 3–4 months old, and some cheese exported from Switzerland may be held for almost a year to acquire a more pronounced flavour.

The composition of a typical Emmental might be:

Protein	28%
Fat	29%
Fat-in-dry-matter	46% (min. 45%)
Moisture	36% (max. 40%)
Salt	1.0–1.6%

The Danish cheese, Svenbo (Figure 5.13), is interesting in that it shows the same expansive openings, and much care is taken to ensure the full activity of the propionibacteria during maturation.

Figure 5.13 Svenbo, a flat, cylindrical cheese coated with cheese wax. The body is firm and sliceable, and the large, round holes are a dominant feature. (Courtesy of the Danish Dairy Board, Aarhus, Denmark.)

Fontal

This variety originates from various areas of Northern Italy, where it is made from full-cream cow's milk. The name, and also that of its closely related variety, Fontina, is derived from 'fondere' (to melt), and it is popular as a 'cooking' cheese. However, to suggest that its use in the kitchen is the usual avenue for exploitation does the variety a disservice, for the mild, delicate flavour makes Fontal an excellent choice as a table cheese.

The method of manufacture is similar to that employed for Gruyère, but the finished cheese (Figure 5.14) is much smaller (diameter 40–45 cm, height 8–10 cm) and the rind has a distinctive dark brown coating. The pale yellow interior is firm and lacking in mechanical fissures, but a few small eyes are always distributed across the cut surface. It has a short maturation period of 30–50 days from manufacture and, in order to prevent moisture loss or loss of quality, the finished cheeses may well be given a coating of paraffin wax. Recently, Fontal has become a registered variety.

The expected composition might be:

Protein	25%
Fat	25%
Fat-in-dry-matter	45%
Moisture	45%

Fontina

Fontina is perhaps the best known of the Italian varieties with openings produced by heterofermentative bacteria, and it has a long tradition of production in the Valle d'Aosta. The granting of Appellation d'Origine status in 1955 has meant that standardization of the characteristics and the method of manufacture are now controlled by a consortium of producers.

Although sheep's milk was often used to make Fontina in the past, raw cow's milk has become the recognized base material and any acidification of the milk relies on a natural flora of mesophilic and thermophilic bacteria. As with Fontal, the initial steps in the production are much like those for Gruyère, and a scalding temperature of 46–48 °C affects both the microflora and the texture of the curd. After moulding and pressing, the cheeses are dry salted initially, but later the rind is kept soft by treatment with brine. This treatment, together with holding at 10–12 °C and up to 90% humidity, encourages the development of *Brevibacterium linens* on the surface of the cheese – a growth which both enhances the flavour and accounts for the typical light orange–brown colour of the rind.

After maturation for around 3 months, the texture will have become soft or semi-hard, and the cut surface will reveal the presence of small round eyes. The flavour is mild and delicate, sometimes almost creamy in nature, and the cheese melts easily and evenly during cooking. In physical appearance, Fontina is a flat, round cheese (40–45 cm in diameter and 7–10 cm in height), and the edges are characterized by being concave. The trademark shown in Figure 5.15 is reserved for authentic products of the Valle d'Aosta and, with production of only around 3000 tonnes per annum, it is a feature that is much valued. Compositional standards are in force as well; the moisture content is specified as 37–40% and the minimum FDM value is 45%. Traditionally, some of the product from sheep's milk would have been stored and used for grating, but nowadays table or culinary uses are the important outlets.

Gouda

As with many Dutch cheeses, Gouda is named after the place in Holland where it was first produced. In many respects, it is much like Edam, but while Edam is usually round, Gouda is always more wheel-shaped, even when produced as the small cheeses (4 kg) shown in Figure 5.16. The large forms – perhaps up to 20 kg – reveal this characteristic even more clearly.

About 60% of the cheese manufactured in Holland is Gouda. Much is produced in modern factories but limited quantities of 'farmhouse' Gouda are available and can be identified by a special designation – 'Boerenkaas' – which is stamped on all such cheeses.

Figure 5.14 Fontal, a medium-hard cheese produced in various regions of northern Italy from cow's milk. (Courtesy of the Ministero dell'Agricoltura e delle Foreste, Rome.)

Figure 5.15 Fontina, a medium-hard cheese produced from cow's milk. (Courtesy of the Assessorato deli' Agricoltura, Foreste e Ambiente Naturale, Regione Autonoma Valla d'Aosta, Italy.)

Figure 5.16 The small, wheel-shaped Gouda is available with a range of flavouring ingredients such as cumin, herbs and pepper. (Courtesy of the Dutch Dairy Bureau, Leatherhead, Surrey, UK.)

One other difference between Edam and Gouda is that the latter usually has a higher fat content, for full-cream pasteurized milk is used in manufacture rather than semi-skimmed milk. Medium-sized vats of 4000–5000 litres are employed in most factories. To the warm milk (30 °C) is added a low level (0.5%) of a starter culture composed of *Lactococcus* spp. and heterofermentative *Leuconostoc* spp. Some propionibacteria may be present as well, because a mixture of small and large eyes is not regarded as undesirable in Gouda; in Edam, only a small number of regular openings is expected (Chapman and Sharpe, 1990). Calcium chloride and sodium nitrate may be added by some producers, and the natural colouring agents, annatto or carotene, can be used to give the curd a more distinctive colour. Given the low level of inoculum, a ripening period may be allowed and, once acid development has begun, around 30 ml of rennet/100 litres of milk should produce a firm gel in 20–30 minutes. This is followed by cutting into small pieces – many with sides as small as 0.5 cm – and the curd/whey mass is stirred for 20–30 minutes. Roughly half the whey is then drained out of the vat, and hot water is added to raise the temperature of the vat contents to 40 °C. Some producers favour a slightly higher temperature to give cheeses with a firmer structure and better keeping quality.

After being stirred at this temperature for around 30 minutes, the curd is allowed to settle and then pressed against the bottom of the vat by means of weighted trays. Once the curd has coalesced, it is cut into convenient blocks and transferred to moulds for pressing. The latter pressure causes the curd to compress into a cheese free from mechanical fissures and, after perhaps 5–6 hours, the cheeses can be removed from the moulds and immersed in saturated brine at 12–15 °C. As with all cheeses that are salted by brining, maintaining the concentration of the solution is critical for success. The cheeses are removed 3–5 days later – depending upon the weight of the individual cheeses – and allowed to drain/dry.

The next stage is waxing to reveal the natural colour of the cheese or with a coloured wax to

Figure 5.17 Around 60% of the total Dutch cheese production consists of Gouda. The young cheese has a soft texture with only a few small, round 'eyes'; as it matures, so the texture becomes drier. (Courtesy of the Dutch Dairy Bureau, Leatherhead, Surrey, UK.)

indicate the addition of cumin or other herbs (Figure 5.16). The cheeses are then matured for 2–3 months at 15 °C for the development of both eyes and flavour. A more pronounced flavour may be achieved by extending the maturation time to 6 months, and some 'mature' brands may be held for up to 2 years before sale. The texture of a young cheese tends to be quite soft (Figure 5.17); as it ages the body becomes firmer and drier, but still easy to cut with a knife. Small, regular openings are clearly visible on the cut surface. A typical analysis of the retail product is shown in Table 5.2.

Table 5.2 Typical composition (%) of Gouda

	6 weeks	6 months
Protein	25	27.5
Fat	30	33
Fat-in-dry-matter	50 (min. 48)	50 (min. 48)
Moisture	40	34
Salt	2.0	2.3

Gruyère

Gruyère is regarded as an important Swiss and French cheese, not least because many English-speaking communities tend to call any cheese with eyes 'Gruyère'. In practice, of course, it is a quite specific variety, which differs from Emmental in:

- being a much smaller cheese in both weight and dimensions;
- having a slightly greasy coat;
- revealing fewer holes of limited diameter on the cut surface (Figure 5.18); and
- tending to have a stronger flavour and aroma.

As with Emmental, cow's milk is the base material but, as no attempt is made to standardize the fat content, the FDM content of Gruyère may be up to 2% higher. After pasteurization, a thermophilic starter culture is added to the cooled milk (30–32 °C), along with propionibacteria, and the rising acidity ensures that the rennet will give a gel in around 30 minutes. The curd is then cut into cubes with sides of up to 1.0 cm, and the stirred curd is then scalded at temperatures which may reach 57 °C. Prolonged holding at this elevated temperature tends to produce quite tough particles of curd, which may be lifted from the vat in the manner of Emmental or, with very large vats, sieved from the whey as it is drawn off. Either way, the curds are collected into moulds and subjected to increasing pressure for 2–3 days to allow time for the curd particles to fuse into a homogeneous cheese; this procedure produces a finished cheese with well-defined, sharp edges and a smooth rind.

Dry salting followed by immersion in brine for 3–5 days at 10–12 °C precedes an initial ripening period of 2–3 weeks at 10 °C, and a longer period of 2–3 months at 15–18 °C to allow for the formation of the eyes. During this latter stage, the cheeses are turned frequently and rubbed with a cloth soaked in brine. This action, combined with holding in a constant humidity of 90–95%, encourages various yeasts and bacteria to grow over the rind, and *Brevibacterium linens*, in particular, contributes to the stronger flavour and

Figure 5.18 Gruyère, a flat, cylindrical cheese (around 40 kg), with a firm texture but slightly pliable; the holes are few and small in diameter. (Courtesy of 'Cheeses from Switzerland'.)

aroma that differentiates Gruyère from Emmental. Final storage for up to 12 months further consolidates this difference, so that the retail cheese has a mature, sharp flavour as against the mild, sweetish notes associated with Emmental.

A typical composition for Gruyère might be:

Protein	30%
Fat	34%
Fat-in-dry-matter	49% (min. 45%)
Moisture	31% (max. 38%)

A variant of Gruyère is Fribourgeois, produced only in the region of Fribourg in Switzerland. It is a much smaller cheese than Gruyère – perhaps no more than 25 cm in diameter and 8 cm in height, compared with up to 60 cm in diameter and 25 cm in height – and tends to be pale in appearance (Figure 5.19). A high FDM level of over 50% makes for a soft, waxy texture, a feature which complements the mild flavour.

Another similar cheese, of Swiss origin, is Raclette (Figure 5.20), which is manufactured in small-scale operations using clean, raw milk.

Figure 5.19 Fribourgeois (Fribourg vacherin), a Gruyère-type cheese with small and infrequent 'eyes'. (Courtesy of 'Cheeses from Switzerland'.)

Figure 5.20 Raclette, a semi-hard cheese that is not widely known outside Switzerland. (Courtesy of 'Cheeses from Switzerland'.)

Figure 5.21 Kefalograviera, a hard cheese made in Greece from a mixture (50/50) of sheep's and cow's milk; the numerous 'eyes' are characteristic of the product. (Courtesy of Professor E.M. Anifantakis.)

Sufficient acidity is developed by the natural microflora of the milk to allow the rennet to act, and a lower scalding temperature than for Gruyère allows a high enough number of naturally-occurring propionibacteria to survive and produce the limited number of eyes required. Maturation at 14–15 °C for about 3 months gives a cheese with the flavour and texture characteristics expected by the consumer.

Kefalograviera

Kefalograviera is a Greek cheese of recent origin, and around 3000 tonnes are produced annually. It is unusual in being produced from a mixture of cow's and sheep's milk (60/40), with occasional inputs from goat's milk (20% maximum). The final cheese is hard with many small eyes scattered over the cut surface (Figure 5.21), and the flavour can be quite strong for some palates.

During manufacture, the mixture of milks is standardized to a fat content of 3.3% (Anifantakis, 1991), pasteurized at 71–72 °C for 15 seconds and then cooled to 33 °C. Calcium chloride and an active mesophilic starter culture are then added. After a short period of ripening, a mixture of rennets is stirred into the acidified milk; a proportion of traditional rennet is usually included, so that the lipases and other enzymes present will enrich the flavour of the end-product. Coagulation will take around 30 minutes, and the gel can then be cut into pieces about the size of maize grains. Following a period of gentle agitation, the temperature of the vat is raised to 47–48 °C over an interval of 25 minutes, and the curds are stirred for a further period until they have become quite firm.

Once the curds have settled to the bottom of the vat, portions sufficient to fill an individual, cloth-lined mould (32 cm diameter × 12 cm high) are removed. Each filled mould is then pressed for about 20 minutes, at which point the cloth is changed, the cheese inverted, and pressing continued for a further hour. This operation is repeated once more before the cloth is finally removed and the cheese is placed in a cool room (12–14 °C) for 24 hours. The cheeses are then placed in tanks of brine (18–20% at 12–14 °C) for two days, before being dry salted (20–25 additions in all) over a period of several weeks. A final wash in brine finishes the salting operation and, after allowing the rind to dry thoroughly, the cheeses are film-wrapped and stored at 8–10 °C for 3 months.

The composition of a typical sample of Kefalograviera might be:

Protein	26%
Fat	31%
Fat-in-dry-matter	48%
Moisture	35%
Salt	3.4%

Maasdam

One of the notable features about Kefalograviera is that the manufacturing procedure was derived by arranging age-old techniques into a sequence that gave an entirely new product, and Maasdam merits attention for much the same reason. Thus, although based on traditional Dutch technology, this relative newcomer to the market is quite distinctive. It is available in two sizes (12 or 16 kg), and the domed surface of the cheese (Figure 5.22) and the massive, regular openings resulting from the action of propionibacteria are unique to this variety. The depth of colour is also unusual for Dutch cheeses.

A typical composition might be:

Protein	30%
Fat	28%
Fat-in-dry-matter	47%
Moisture	40%
Salt	1.4%

Figure 5.22 Maasdam, a flat, cylindrical cheese with a thin coat of cheese wax. The noticeable 'doming' and the large holes are distinctive features of the variety. (Courtesy of the Dutch Dairy Bureau, Leatherhead, Surrey, UK.)

Whether the low salt content is a reaction to consumer concerns or is employed to avoid any depression of the metabolism of the propionibacteria is not clear, but gas production is certainly encouraged at low salt levels. The flavour of the cheese is reported to be 'mild and nutty', and it will be interesting to see how the market responds.

Maribo

This Danish cheese is alleged to have been derived from Gouda, but an examination of the cut surface (Figure 5.23) reveals a mixture of eyes and mechanical fissures that is entirely different from the Swiss variety.

Pasteurized cow's milk standardized to 3% fat is employed for manufacture. According to Davis (1976), buttermilk and sodium nitrate may be added prior to cooling the milk to 30 °C, and the nitrate will, as usual, be expected to suppress any spoilage by undesirable bacteria. The buttermilk, however, will not only serve as an additional source of protein but also its content of diacetyl

may enrich the flavour of the finished cheese. A mixed mesophilic starter culture is added along with around 30 ml of rennet/100 litres of milk, and the coagulum should be ready for cutting some 50–60 minutes later. The gel is cut into cubes with sides of approximately 1 cm, and the curd/whey mass is stirred for about 30 minutes. This process is then continued as some 30–35% of the whey is drawn off and sufficient hot water added to raise the temperature of the contents of the vat to 37–38 °C. A prolonged holding and stirring at this latter temperature follows, until finally the whey is drained completely and the warm curds are mixed and salted. Portions of the curds are then moulded and pressed for around 30 minutes. At this point, the rough-formed cheese is wrapped in cloth, inverted and subjected to more severe and prolonged pressure.

After several hours, the cheeses are removed from the press and allowed to dry overnight at room temperature. Over the next 2–3 days, they are immersed in brine (24% salt) and then, after draining and surface drying, they are transferred to a ripening room at 17–18°C. Four weeks

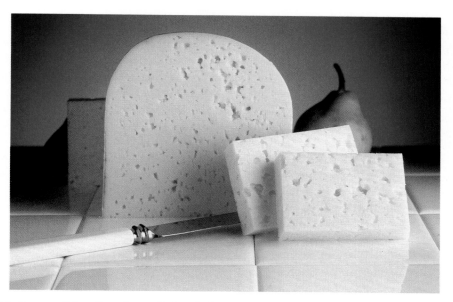

Figure 5.23 Maribo, traditionally a flat, cylindrical cheese coated with yellow cheese wax. The body is firm and sliceable, and the cut surface shows numerous irregular, small holes. (Courtesy of the Danish Dairy Board, Aarhus, Denmark.)

under these conditions allows the heterofermentative starter bacteria to generate numerous small gas holes in the body of the cheese, and these, along with natural fractures, give the cut cheese its characteristic appearance. A further period of storage at a lower temperature completes the maturation, and by maintaining low humidities (never above 85%) throughout the sequence, a hard, dry rind develops that may then be coated with yellow cheese wax for ease of handling.

Although block forms are now available, the traditional cheese is a flat cylinder (40 cm in diameter × 10 cm high) weighing about 14 kg. The flavour is rich and slightly acidic, and the minimum FDM level is 45%; the maximum moisture content is set at 43%. A variety with caraway seeds is available but its production is not widespread.

Samsoe

Samsoe is one of a family of Swiss-style cheeses that are produced in Denmark, and the differences between them, at least in terms of flavour and texture, tend to be subtle rather than obvious. Similarly, contrasts between the processes tend to reflect the introduction of seemingly quite minor adjustments, and this situation is apparent with the manufacture of two cheeses, Samsoe (Figure 5.24) and Fynbo (Figure 5.25). The outward appearance of the two varieties is totally different but the production systems have much in common.

The milk is standardized to a level demanded by the regulations covering FDM, and then pasteurized at 72 °C for 15 seconds (Anon., 1992) and cooled to 30 °C. A mesophilic starter culture, perhaps with additional propionibacteria, is added to the vat, and sodium nitrate is frequently added to control any unwanted gas formation by contaminant bacteria. Once the acidification of the milk has begun, around 30 ml of rennet/100 litres of milk is distributed evenly throughout the vat so that a firm gel forms in 30–40 minutes. Cutting into 5 mm cubes follows, and the curd/whey mixture is stirred for 30–35 minutes. Around 30% of the whey is then removed, and replaced with water at 65–67 °C. This addition brings the temperature of the vat contents to 35–37 °C (at the lower extreme for Fynbo); the curd is then stirred for around 40 minutes for Fynbo and rather longer for Samsoe. A low level of salt is sometimes added during the mixing, presumably to assist in firming up the particles of curd.

As with many compact cheeses, the initial pressing of the curd takes place beneath the whey at the bottom of the vat, and only when the curd has fused together without air-spaces is the whey run out. The warm curd is then transferred to moulds of the dimension required for the finished cheese (44 cm in diameter × 10 cm high for Samsoe, but of various sizes for Fynbo/Mini Fynbo) and pressed for several hours at ambient temperatures of up to 20 °C. The cheeses are removed from the moulds for the rind to dry overnight at 12–14 °C, and then immersed in brine (24–25% salt) for 3 days.

The initial ripening, which may take 4 weeks at 18–20 °C and 83–85% RH, allows time for the small, regular openings to develop, and a further period at lower temperatures and humidities encourages flavour development. The cheeses are then waxed and stamped in accordance with local custom and current regulations, and matured for 3–5 months depending upon whether the cheese is for local consumption or for export. Over this time, Samsoe takes on a firm texture and rich, nutty flavour, but Fynbo tends to remain rather mild. Both cheeses appear white to pale yellow on cutting, and display a limited number of regular eyes. The minimum FDM for both Samsoe and Fynbo is set at 45%, although a low-fat version of Samsoe is available. The maximum moisture contents are 44% for Samsoe and 48% for Fynbo.

Scandinavian cheeses

Both Norway and Sweden produce a number of cheeses in the style of Gouda but, because production is on a limited scale, international recognition tends to be limited. Only travellers to

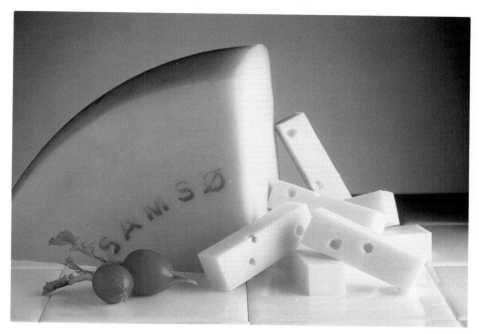

Figure 5.24 Samsoe, a flat, cylindrical cheese coated with cheese wax. The body is firm and sliceable; occasional round holes are an essential feature. (Courtesy of the Danish Dairy Board, Aarhus, Denmark.)

Figure 5.25 Mini Fynbo, a pale cheese coated with red cheese wax. The body is firm and sliceable, and the regular, round holes are limited in number. (Courtesy of the Danish Dairy Board, Aarhus, Denmark.)

Sweden, for example, are likely to have encountered Sveciaost – a Gouda-style cheese with numerous irregular openings instead of the round eyes associated with gas production – or the hard, long-matured Vasterbottenost. The latter cheese again has more irregular fissures than true eyes because unlike many of the Danish cheeses discussed earlier, the initial pressing of the curd takes place in the mould rather than in the vat under whey; consequently, natural air-pockets remain in the curd and, even though propionibacteria may be active, the gas does not become trapped to any real extent.

In Norway, cheeses more typical of the Swiss varieties are available, and one on the market under the brand name Jarlsberg (Figure 5.26), could well prove popular with consumers across Europe. The normal cheese is around 33 cm in diameter, 10 cm high and 10 kg in weight; it has a smooth, sliceable texture broken by relatively large eyes. Its rich, almost sweet, flavour is typical of cheeses manufactured with propionibacteria in the culture.

A typical analysis might be:

Protein	27%
Fat	27%
Fat-in-dry-matter	45% (min.)
Moisture	40%
Salt	1.5%

Figure 5.26 Jarlsberg, a soft-textured cheese with a distinctive 'nutty' flavour and an abundance of large 'eyes'. (Courtesy of the Norwegian Dairies Association.)

6 Cheeses ripened with moulds

The popular cheeses in this group fall into two broad categories: those characterized by the presence of a mould growing as a visible 'felt' over the surface of the cheese, and those in which the mould is restricted to fissures within the structure of the cheese. In both cases, species of *Penicillium* are involved as the dominant fungi, and the essential rôle of these moulds is to enhance the consumer appeal of the cheese by modifying the normal processes of maturation. However, in spite of these unifying features, the manufacture of the two groups follows very different pathways, and hence it is convenient to consider the different varieties within their specific subgroup.

Cheeses surface-ripened with fungi

These varieties of cheese have much in common with the soft, unripened varieties that are found in many regions of France. Coulommier is a case in point, and the traditional system of manufacture could well have provided the basis for cheeses like Brie and Camembert.

In essence, cow's milk is pasteurized and cooled to around 30 °C, prior to inoculation with a low level of a starter culture of acid-producing lactic acid bacteria. Rennet (0.3% v/v) is added and the milk is allowed to stand for up to 2 hours. During this time, the milk coagulates to form a firm gel, which can then be ladled into two-part cylindrical moulds of some 10 cm in diameter and with a total height of around 20 cm. Standing on a rush or bamboo draining-mat, the entire mould is filled with curd. After 24 hours at ambient temperature, the curd will have drained and contracted so that only the bottom half of the mould is occupied, and the top section can be removed.

The lower section, including the curd, is now inverted on to a clean draining-mat so that the remaining whey can exude from the cheese. Some 24 hours later, the finished cheese can be removed from the mould and the exposed surfaces coated with dry salt. The salting, together with the drying action of exposure to the atmosphere, serves to harden the surface of the cheese and it is then ready for immediate consumption. Some varieties, particularly if they have been manufactured with a culture including aroma-producing bacteria, may be matured for 1–2 weeks at 10–12 °C prior to sale but, either way, Coulommier has an extremely restricted shelf-life.

Systems along these lines evolved world-wide, and the natural susceptibility of the products to mould contamination and growth led to the establishment of many local varieties. Some types have gained official recognition and are now produced in countries far removed from their original location, whilst others have remained of local interest only. Consequently the following coverage has, in the main, been restricted to varieties of international interest.

Anthotiros

Anthotiros is more commonly sold as a fresh cheese nowadays, but as the flavour of the traditional product was dependent on the growth of external moulds, it is not unreasonable to place it in the present category.

It is produced on the island of Crete from the whey released in the manufacture of Kefalotiri cheese. A small quantity of whole sheep's or goat's milk is added to increase the total solids. The system of production is very similar to Mizithra, but the higher fat content of Anthotiros – associated with the local milk – is reported to give the Cretan variety more pleasant organoleptic characteristics.

The fresh cheeses (Figure 6.1) may be consumed directly, often with honey, or they may be dry salted and hung in nets in cool, well-ventilated rooms. A variety of yeasts and moulds colonize the moist surface of the cheeses but, as the moisture content of the curd falls, the surface dries and microbial activity ceases. A final moisture level of 40% is common, and the mature product may be used as a table cheese or crumbled over salads or other dishes.

Brie

The Brie of normal commerce is a large, circular, thin cheese covered with a surface layer of white mould, *Penicillium candidum*. It develops a distinctive, piquant flavour with age, but a young cheese has a mild, slightly aromatic taste. The texture is close and smooth and, although usually soft and spreadable depending upon the degree of maturation, it can sometimes be quite crumbly.

The basic type is now manufactured in many countries, but in its country of origin, France, a number of unique varieties have come to be recognized, e.g. Brie de Coulommier, Brie de Melun and Brie de Meaux (Figure 6.2). The existence of this range reflects the fact that, from its origins sometime before the 12th century and until quite recently, Brie was essentially a farmhouse cheese. As a result, many totally original cheeses emerged in or around the Île de France region.

Traditionally, fresh morning milk, together perhaps with some cooled milk from the previous evening, is poured into vats (100 litres) together with rennet and a starter culture of *Lact. lactis* and *Lact. lactis* subsp. *cremoris*. At a temperature of 30–31 °C, a firm coagulum is formed in 2–3 hours, and this gel is then cut with a coarse 'wire' knife into slices some 3 cm thick. Segments of these 'slices' are then carefully transferred into metal hoops of the desired dimensions, laid upon a draining-table covered with rush or nylon mats; alternatively, the gel can be simply ladled into the

Figure 6.1 Anthotiros, a traditional whey cheese that may be consumed fresh or used as a condiment. (Courtesy of Professor E.M. Anifantakis.)

Figure 6.2 Brie de Meaux, often regarded as the classic form; the large diameter, shallow depth and soft texture necessitate the secure packaging of each individual cheese. (Courtesy of Texel, Epernon.)

hoops without any previous cutting. Either way, the important feature of the hoops is that they are manufactured as two interlocking rings which stack one on top of the other, and the entire two-ring structure has to be filled with the soft gel.

Over the next 18–24 hours, the whey drains out leaving a curd that occupies the lower section of the hoop only. The upper section is then removed, and the cheese is turned over on to a clean mat. Sometimes this turning process is repeated to facilitate drainage. Once firm, the exposed surfaces are dry salted – coarse salt is used for the first dressing, but fine salt can be employed for any subsequent treatments. The advantage of coarse salt is that it dissolves more slowly than the fine material, and hence does not cause the surface of the cheese to harden and trap whey that would otherwise drain out. The salt may also act as a carrier for the spores of the *Penicillium camemberti* (*P. candidum*), but better distribution of the spores is often attained by incorporating them into the milk along with the rennet and the starter. After an initial holding period in a well-ventilated room at 14 °C to allow rapid development of the mould, the cheeses –

minus the constraining hoops – are transferred to a ripening store at 10–12 °C and 85% RH.

This mould has an active and varied metabolism, and of especial relevance is its proteolytic activity. Both proteases and peptidases are released into the cheese, and it is the action of these enzymes on the caseins in the curd that results in the characteristic softening. The release of lipases into the curd is equally important, as is the ability of the mould to metabolize the fatty acids (liberated by the breakdown of the milk fat) to flavour-enhancing components. The extent to which these changes occur has a profound influence on the flavour and aroma of the finished cheese, as may the concomitant growth of adventitious bacteria, yeasts or moulds, such as *Geotrichum candidum*. These latter organisms are natural inhabitants of the cheese room, and often develop on the surface of the cheese – a growth assisted by the ability of *Penicillium* to metabolize the lactic acid and reduce the acidity of the curd. The nature of this casual microflora and its impact on flavour and aroma cannot be specified, but the overall result will be to increase the complexity of the sensory qualities and perhaps provide cheeses from a given region or factory with a distinctive character.

Although a traditional farmhouse variety, most Brie is now made in large factories around Paris or in eastern France (Lenoir and Tourneur, 1993). The finished cheese should contain at least 40% FDM – 45% in the case of Brie de Meaux – and the moisture content will be in the region of 56%. The common commercial sizes are: Brie de Meaux around 36 cm diameter, the dimensions of which make it most suitable for subdivision into retail portions; and the smaller Brie de Coulommiers approximately 14 cm diameter. Of the two, Brie de Meaux is perhaps the more typical of the variety and, as such, enjoys an Official Denomination of Origin. It must be made in the traditional fashion, and a 4-week ripening period must take place in a region adjacent to the rivers Seine or Marne. The tonnage produced in this way is comparatively small (6000–7000 tonnes per annum), and much Brie on the market is manufactured under very different conditions.

Figure 6.3 Different types of Brie are now manufactured in many countries, and this variant from Germany contains peppercorns. (Courtesy of CMA (UK), London.)

Special varieties are also available and the example shown in Figure 6.3 is typical of the modifications that have taken place world-wide.

Camembert

According to Lenoir and Tourneur (1993), Camembert was first produced in Normandy around 1790. Traditionally, it is a small cheese with a diameter of around 10 cm, a moisture content of 46–47% and an FDM of at least 40%.

The Official Label of Origin is applied to Camembert de Normandie, which is made from raw milk produced in designated regions in Normandy. A low level of starter culture is added – perhaps 0.1% of a mesophilic culture – and once the pH has fallen to around 6.1 a commercial coagulant is added at a rate of around 20 ml/100 litres of milk. At 30 °C, the milk will coagulate in some 1–1½ hours to give a coagulum of sufficient firmness for ladling into moulds rather in the manner of Brie. Overnight drainage

provides a cheese capable of withstanding extraction and covering with a mixture of dry salt and spores of *Penicillium camemberti*.

After surface-drying at ambient temperature in a well-ventilated room, the cheeses are transferred to a cool ripening room (12 °C) for the mould to grow over the surface of the cheese. The correct balance between aeration and humidity is essential to ensure rapid development of the mould but, unlike the situation with Brie, the fungus may not remain the dominant member of the microflora. Thus, although a thick felt of fungus may cover the cheese after 10–12 days, high population levels of bacteria, yeasts and *Geotrichum candidum* gradually emerge to modify the coat of the finished cheese. This complex microflora initiates many changes in the curd, particularly under the influence of lipolytic and proteolytic enzymes, and it is the resultant changes in texture and flavour that give traditional Camembert its unique character.

The final stage is packing in waxed paper and the distinctive wooden boxes, and final flavour development takes place at around 7 °C over the next 7–10 days. Thus, full ripening of the hard, granular curd through to the creamy-yellow, pliable cheese (Figure 6.4) may take around one month. With production of only around 10 000 tonnes per annum, the product is highly prized by connoisseurs.

For many markets, however, Camembert of industrial origin is perfectly acceptable and France produces in the region of 130 000 tonnes per annum in large automated factories. The essential differences are that:

- the bulk milk is pasteurized;
- the spores of *Penicillium camemberti* may be added to the milk before coagulation, rather than during salting, to ensure an even distribution within the final cheese;
- the coagulum, which may be formed more rapidly by the use of higher temperatures and addition rates for the rennet, is cut mechanically into 2.5 cm cubes prior to filling into moulds; and
- brining may, for ease of automation, replace dry salting.

Figure 6.4 Camembert, one of the classic cheeses of Normandy, with origins in the 12th century; it is now produced widely around the world. (Courtesy of the CMA, Bonn, Germany.)

Maturation does, of course, still rely on the activity of the surface microflora, but the bulk handling of tonnage quantities allows the manufacturer to achieve the high through-put necessary to satisfy the mass market.

Carré de l'Est

As the name suggests, this cheese has its origins in the east of France and it should always be square in shape. Beyond these superficial points, it resembles Brie or Camembert, although the flavour is normally milder than Camembert. The modernization of the process has tended to follow the other cheeses also, for automation has now become essential for the economic survival of much of the cheese industry.

Some small plants do still follow tradition, and

the conventional process involves standardizing cow's milk to around 3.2% fat and heat treating at 72 °C for 15 seconds. After the milk has been cooled to 32–33 °C, a standard mesophilic starter culture (1%) is added, and the acidity of the milk is allowed to develop for around one hour. A comparatively low level of rennet (15–20 ml/100 litres of milk) produces a firm coagulum in 50–60 minutes. This gel is cut coarsely, then gently stirred and transferred to square moulds on draining-mats. As the curd coalesces to form a cheese, the moulds are turned to facilitate free drainage of the whey.

Next morning, the cheeses, which should be quite acidic at this point (pH 4.8), are removed from their moulds and covered with a mixture of salt and spores of *P. camemberti*. A holding period of 24 hours at 18–20 °C allows the spores to adhere firmly to the curd, and the cheeses are then surface-dried for 48 hours at 14 °C before being placed in a cool room (11–12 °C) for the mycelial coat to develop over a period of around 2 weeks. The main changes introduced in recent times are geared to speeding up the process:

- An increased level of starter culture is employed, along with an elevated ripening temperature.
- The mould spores will be incorporated into the milk.
- The curd/whey mixture will be pumped into moulds, and the subsequent turning operations will be mechanized.
- The finished cheeses will be brined rather than dry salted, again opening the way for automated handling.

Irrespective of the precise system of manufacture, the retail cheeses are around 10 cm square and 2.5 cm thick (Figure 6.5), and a typical cheese should have a moisture content of 50–53% and an FDM of 45–50%.

A number of similar varieties have originated in the same region of France. Chaource (Figure 6.6) is a typical example of a cheese which has become important to the economy of the region, but limited supplies have reduced its impact on wider markets.

Figure 6.5 Carré de l'Est, a square cheese from the Alsace that has much in common with Camembert. (Courtesy of CIDIL, Paris.)

Figure 6.6 Chaource, a thick, Brie-type cheese produced in limited quantities in the Champagne region of France. (Courtesy of CIDIL, Paris.)

Goat's milk cheeses

There are so many goat's milk cheeses produced world-wide that classification is wellnigh impossible, the more so as differences between so-called varieties are often both negligible and subject to much natural variability. The manufacture of a number of fresh cheeses has become more standardized at the behest of supermarket chains, but those that involve mould-ripening and maturation remain primarily for local consumption. This pattern is particularly well defined in France, for example, and many soft cheeses of various shapes can be found at local markets; in the Poitou/Limousin region alone, over 30 different varieties of goat's milk cheese have been named (Stobbs, 1984). The flavour of such cheeses is usually strong, and the heavy presence of short-chain fatty acids tends to give the products a characteristic flavour. The rinds are often coloured as a result of mould growth.

A typical example is Chabichou (Figure 6.7), a small, soft cheese (100 g) with around 45% FDM. The truncated cone shape is typical of

Figure 6.7 Chabichou, a quick-ripening goat's milk cheese produced for local sale around Poitou. (Courtesy of CIDIL, Paris.)

many goat's milk cheeses, and Pouligney-St Pierre from the Pays de la Loire region (Figure 6.8) illustrates just how widespread this configuration has become over the years. On removal from the mould and dry salting, the exposure of such cheeses at room temperature allows a range of penicillia to grow over the surface, and during the summer months when they are usually produced, the surfaces of the cheeses rapidly become green-grey/brown in colour. By the end of 3 weeks, the cheese is ready for eating. If it is left any longer, the colour of the rind darkens and the flavour becomes markedly stronger; the loss of moisture tends to highlight the salt content as well.

Although manufacture on the same farm will, over the years, tend to select for a determinate range of moulds, the lack of control over the microflora does lead to much variability within and between batches. Visible evidence of the casual nature of this mould growth can be seen at early stages of maturation, and Figure 6.9 shows a mixture of white and blue moulds on the surface of Crottin de Chavignol, a goat's milk

Figure 6.8 Pouligny-St Pierre, a soft goat's milk cheese with a strong aroma and taste. (Courtesy of CIDIL, Paris.)

Figure 6.9 Crottin de Chavignol, a soft goat's milk cheese of variable appearance and quality, often produced on the farm. (Courtesy of CIDIL, Paris.)

Figure 6.10 St Maure, a strong goat's milk cheese with the characteristic shape of an elongated cylinder. (Courtesy of CIDIL, Paris.)

cheese from Pays de la Loire. The uncontrolled nature of this microflora gives rise to end-products which are equally variable, differences that may well be exaggerated by the maturation time of up to 3 months. At the end of this time the cheese will have dried, the coat becomes pale brown in colour and the flavour too strong for many palates. This latter property may explain why much of the cheese is sold direct from farms, for in this way the unwary will not be disappointed by an impulse purchase.

While truncated cones have a certain aesthetic appeal, the log-shape characteristic of St Maure offers practical advantages for the consumer wishing to remove a generous portion (Figure 6.10). It is produced in both the Poitou and Pays de la Loire regions, and the dry, brown rind indicative of earlier mould growth is a common feature. Sold at around one month old, the flavour is usually strong and highly characteristic of goat's milk products.

Neufchâtel

This mould-ripened cheese originated in the area around Rouen and, like many such cheeses, much product is still made on the farm. A small quantity of home-prepared starter is added to raw cow's milk – sometimes partially skimmed – at 30 °C, along with a low level of rennet, and the milk is then allowed to stand overnight. Next morning, the curd is drained by hanging in a cloth bag and later may be cooled and pressed under weighted boards. When most of the free whey has been lost, the coarse curd is mixed to a more homogeneous consistency and salted, prior to being transferred into moulds. Once the curd has taken the shape of the mould, it is removed and dusted with a mixture of salt and spores of *P. camemberti*. The surface of the cheese is then allowed to dry at room temperature for several hours, at which point the cheeses are moved to a cool room for the white coat of mould to develop. Around 2–3 weeks later, the cheese is ready for consumption.

Although the moisture content can be as high as 60% or more for the fresh cheese, most Neufchâtel will analyse as 51–53%, and give an FDM value of 45–51%. The shape and dimensions of the cheeses are mainly at the discretion of the producer (Figure 6.11) or are typical of a given region – as are the salt contents, which may range from 0.8% to 1.5%.

Figure 6.11 Neufchâtel, one of the traditional cheeses of Normandy, alleged to have been made as long ago as AD 1050. A number of variants are produced and this heart-shaped form is known as 'Coeur-de-Bray', after the Pays-de-Bray region near Rouen. (Courtesy of CIDIL, Paris.)

Figure 6.12 Tomme de St Marcellin (St Marcellin), made in the Isère Valley from goat's milk or a mixture of cow's, sheep's and goat's milk. The irregular coat of moulds is quite typical. (Courtesy of CIDIL, Paris.)

Tomme de St Marcellin

Although originally made from goat's milk, this cheese evolved to be produced from cow's or sheep's milk or any convenient mixture. According to Stobbs (1984), there is a firm record of the cheese first being named around 1460, and it is probably typical of the numerous local varieties that have emerged across Europe in general, and France in particular.

As shown in Figure 6.12, it is a cheese of 8–10 cm in diameter and 2–3 cm in thickness. Simple exposure to the atmosphere at ambient temperature allows a range of moulds to colonize the surface. Frequent usage of the same rooms leads to the natural selection of a dominant flora, and it is likely that the white/blue moulds visible in Figure 6.12 are penicillia. Some control of the microflora may be exerted in some places by providing a humid atmosphere during the early stages of ripening – perhaps for 2 weeks to ensure complete coverage of the surfaces by the fungus. However, the coat of the retail product is always

dry, and hence at least 2 weeks at low humidity is necessary to enable the rind to dry; full maturation will be achieved in around 4 weeks.

Cheeses internally ripened with moulds

The cheeses in this group are all characterized by having a rather soft curd veined with lines of green or blue where the mould (usually *Penicillium roqueforti*) has developed. The flavour that develops as a result of this fungal growth is both strong and distinctive, and many connoisseurs feel that no cheese-board is complete without a blue-veined variety. The most famous varieties like Roquefort, Stilton and Gorgonzola enjoy long histories of production, but many cheeses, such as Danish Blue and Bleu d'Auvergne, have evolved as major competitors in the market place.

The distinctive flavour derives from a combination of correct processing of the curd to leave a soft, high moisture base through which the mould can ramify, correct aeration of the curd to ensure that the blue veining develops and, of course, the metabolic activity of the mould itself. Establishment of these conditions is as much an 'art' as a 'science', and so temperamental is the maturation process that each individual cheese of a highly-prized variety like Stilton may have to be examined prior to sale. It is demands of this type that tend to make the retail price of some varieties of mould-ripened cheese rather expensive *vis-à-vis* widely consumed types like Cheddar or Feta but, to date, systems for the automation and cost reduction of the manufacturing process have yet to be developed.

Bleu d'Auvergne

This cheese is perhaps one of the best known 'blue cheeses' made in France from cow's milk and it was originally a farmhouse cheese produced in Auvergne. Factory production dominates nowadays, but, even so, annual output does not exceed more than a few thousand tonnes.

Typically, raw milk is employed, and good quality milk is poured into a series of small vats and heated to 30–32 °C. Around 30 ml of standard rennet/100 litres of milk is added, and because acidification relies on the natural microflora, coagulation will take 90 minutes or even longer. Spores of *Penicillium* may be added at this stage as well, but some makers prefer to introduce them as the curds are being transferred to the moulds.

After cutting, the curd is allowed to settle slowly on to the bottom of the vat and whey is drawn off. In the absence of any scalding, the curd is very soft at this point and it is scooped from the vat into cloth-lined drainers to acquire an initial structure. Moulding is not unlike that employed traditionally for Coulommier except that the cylindrical moulds are larger (around 22 cm in diameter and 10 cm high) and perforated. Consequently, compaction of the curds is quite rapid.

Once the upper section has been removed, the cheeses are held at room temperature for 3–4 days and turned at regular intervals. Dry salting follows, and further additions will be made over a period of about one week while the cheeses stand in a cool room at 10 °C or thereabouts. They will remain under these conditions, with occasional turning, for around 3 weeks. The developing rinds are then washed clean with brine, and the cheeses are dried and transferred to a maturing room at 8–10 °C and high humidity. Aeration of the interior of the cheese is encouraged by stabbing with wire needles (see Stilton, below) and the cheese will be ready for sale some 2 months later.

It is a quite salty cheese with a definite, sharp flavour, and the open texture shows clear areas of blue mould development (Figure 6.13). Wrapping in metal foil is favoured to maintain the soft, moist nature of the product (maximum of 50% moisture), and the FDM value will be at least 40%.

A number of other blue-veined cheeses are produced on farms or in small factories across France. A typical example is Bleu des Causses, a flat, cylindrical cheese of around 2 kg, which is made all year round from cow's milk. The irregular fractures are more pronounced than in many such cheeses, and the domination of such areas by the *Penicillium* is quite pronounced (Figure 6.14). Differences in pasture and/or maturation conditions provide certain features that are distinctive, at least to those well-versed in the art of cheese tasting, but the boundaries between many varieties of 'blue cheese' do tend to be a little blurred for the casual consumer.

Cabrales

Raw cow's milk has become the normal source of Cabrales nowadays, but it comes from a mountainous region of Spain and so some local forms are made from goat's or sheep's milk or a mixture of available materials.

The milk is heated to 35 °C, and coagulated with standard rennet; it may take 2–3 hours for

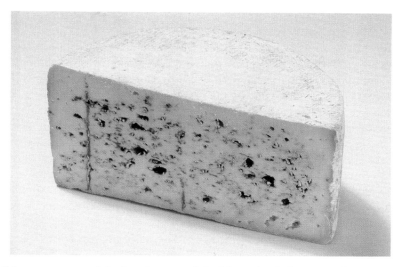

Figure 6.13 Bleu d'Auvergne, one of the best-known of the blue-veined cheeses made in France from cow's milk. (Courtesy of CIDIL, Paris.)

Figure 6.14 Bleu des Causses, a soft blue-veined variety produced from cow's milk in south-west France. (Courtesy of CIDIL, Paris.)

the gel to form in the absence of any starter culture. After cutting into cubes with sides of around 2.0 cm, the curd is allowed to stand for 24 hours at room temperature. The curd is then placed in moulds that will give a finished cheese of 20–30 cm in diameter and 7–15 cm in height. Given this degree of local variation, the weights of the cheeses will range from 1 kg to 5 kg.

Once the curds have fused into a defined shape, the cheeses are dry salted on two or three occasions over the next few days, before being transferred to local caves for maturation. Covering the cheese with leaves of the maple (*Acer pseudoplatanus*) helps to retain a degree of humidity in the caves, and an ambient temperature of 10 °C encourages the development of the mould over the next six months. Natural infections from *Penicillium roqueforti* provide the blue veining that can be seen in Figure 6.15, and because no spores are added to the milk or curd, the degree of mould growth is extremely variable. Now that Cabrales has received an Appellation d'Origine, attempts are being made to standardize both production techniques and chemical composition, even though the annual output is little more than 100 tonnes.

A typical analysis might be:

Protein	22%
Fat	33%
Fat-in-dry-matter	57%
Moisture	42%

Drier, more crumbly cheeses are not uncommon.

Danish Blue cheese (Danablu)

Although a comparative newcomer to the commercial market, Danish Blue is probably the best known of all the so-called 'blue cheeses'. It has gained this recognition as a result of extensive production on a world-wide basis so that, for many people, Danish Blue is the only cheese of this genre that is readily available. A lower retail price *vis-à-vis* Stilton or Roquefort provides another strong incentive for selection, and hence the variety has to some extent come to dominate the market for cheeses internally ripened with moulds.

The basic process is broadly similar to that for Roquefort (see below), but one important difference is that the raw cow's milk is homogenized prior to pasteurization. The effect of this homogenization is to expose more of the fat to the action of lipolytic enzymes released by the mould, so that flavour development is both more rapid and, with increasing age, more intense than with some traditional cheeses. It is claimed also that homogenization assists in the formation of

Figure 6.15 Cabrales, a blue-veined variety made from cow's milk or sometimes other milks in the region of Asturias. (Courtesy of the Ministerio de Agricultura Pesca y Alimentacion, Spain.)

an open-textured cheese (Shaw, 1993) that encourages a luxuriant growth of *Penicillium roqueforti*.

After homogenization, the milk is pasteurized and, in some factories, whitened by the addition of a chlorophyll preparation derived from alfalfa or some similar crop. This green material has the effect of masking, to a limited degree, the yellow hues associated with the carotenes in cow's milk and, as a result, the final curd appears distinctly white. A standard, homofermentative starter culture is employed at 31–32 °C to produce the initial acidity in the milk, and around 30 ml of rennet/100 litres of milk is sufficient to form a coagulum within an hour of addition. After being cut into small cubes with sides of 10–15 mm, the curd/whey mixture is heated sufficiently (perhaps to no more than 33 °C) to cause the curd to become firm, but not to have any dramatic effect on the starter culture. The resultant high population of starter bacteria ensures that acid development continues during the moulding and draining stages, so that at 24 hours the pH may be as low as 4.5.

Following the mild scald, the whey is drained out of the vat and the curd is piled along the sides for salting (2%) and the addition of the spores of *P. roqueforti*. Large cylindrical moulds capable of producing a cheese of around 3 kg are common, but smaller cheeses of some 10 cm in diameter are produced along with 4 kg blocks that are convenient for portioning (Davis, 1976). Once filled, the moulds are held overnight at 23–24 °C for natural drainage of the curd to continue, a process usually assisted by turning the moulded cheese at some point. Dry salting or immersion in brine completes the preparatory stages, and the cheeses are then matured at 10–12 °C for up to 3 months. Extensive piercing encourages a prolific development of the mould and concomitant enzyme activity, so that the finished cheese has an abundant venation of blue-green mould and a strong, pungent flavour. Typical analyses of two popular types of Danish Blue cheese are shown in Table 6.1. The texture should be spreadable/sliceable, and the slightly moist, white or yellowish curd is often broken with irregular holes (Figure 6.16).

Figure 6.16 Danablu (Danish Blue cheese), showing the characteristic white background with greenish-blue veins. The aeration holes made during manufacture are clearly visible. (Courtesy of the Danish Dairy Board, Aarhus, Denmark.)

Table 6.1 Typical composition (%) of Danish Blue

	Mellow	Mature
Protein	20	21
Fat	35	26 (min.)
Fat-in-dry-matter	60	50 (min.)
Moisture	42	48 (max.)
Salt	4.5–5.0	4.5–5.0

The equivalent cheese from Austria and Germany is Edelpilzkäse (Figure 6.17). It is reported to have a slightly milder flavour than Danish Blue – despite the more abundant mould growth – and is a popular table cheese in central Europe.

Figure 6.17 Edelpilzkäse, a cheese found in Austria and Germany, with characteristics that suggest a close relationship with Gorgonzola. (Courtesy of the CMA, Bonn, Germany.)

Gorgonzola

Gorgonzola is the traditional blue-veined cheese of Italy. It takes its name from the village of Gorgonzola near Milan, for it was here that the cheese was first made from the milk of cattle driven into the region to pass the winter. Turning the surplus autumn milk into cheese allowed maturation to take place during the cold months of the year, and hence excessive ripening and spoilt products were avoided. Nowadays, controlled curing systems allow manufacture on a year-round basis in many parts of Italy and, although the name is protected, in other countries as well.

Although it can be manufactured in a manner not unlike Stilton, the classical process involves the production of two separate curds, one from the evening milk and one from the morning milk. The evening milk is processed first and, after warming to 28–32 °C, around 2% of a normal mesophilic starter culture is added. Once the initial acid development has taken place, rennet is added to give a firm coagulum some 20 minutes later. After the curd has been cut into coarse cubes, the curd/whey mixture is allowed to stand, with occasional stirring, for 30 minutes or so to permit further acid development and firming of the curd pieces. Some of the whey is then drained off, and the remaining curds and whey are scooped into cloth bags for draining overnight at ambient temperature.

The morning milk is handled in a roughly similar fashion, but before the drained curd has an opportunity to cool, the moulding stage takes place. Wooden hoops (20–30 cm in diameter and around 25 cm deep) are lined with cloth and placed upon draining-mats, and the warm curd is layered into the base and up the sides of the mould. Spores of *Penicillium roqueforti* – often known as *P. glaucum* in Italy, are sprinkled over the exposed curd, and then mixed with the cooler (evening) curd as it is filled into the centre of the mould. A topping with more of the warm morning curd completes the filling operation. Although complicated, this double-curd approach is considered critical for the production of top quality Gorgonzola, for it both encourages mechanical openings in the centre of the cheese for growth of the *Penicillium*, and gives rise to a cheese with a smooth, firm surface (Figure 6.18).

After the top layer of curd has been added, the cloth is gathered over the surface and the cheese is turned. Frequent turning over the next 24 hours encourages draining of the whey and consolidation of the curd, so that by the next day the cheese is firm enough for the cloth to be removed. The hoop is retained, however, to assist with further turning of the cheese over the next few days. Once the surface has dried sufficiently, dry salting can take place. This rubbing of dry salt on to the surface takes place almost daily for around 2 weeks, and cool conditions (i.e. around 10 °C) are considered ideal. An intermediate ripening stage of one month at 12–14 °C (humidity 75–80%) follows, and frequent turning and cleaning reduces the risk of unwanted surface discolouration.

Transfer to a cooler area of 9–10 °C and higher

humidity is necessary for the second stage of maturation. If the curd has settled too firmly, spiking of the cheese to produce aeration channels may take place. Under these conditions, the *Penicillium* should develop as visible blue–green pockets within the white or straw-coloured curd, and the cheeses are then moved to a cold room at 4 °C for the final period of ripening. A humidity in excess of 90% is essential at this latter stage to avoid splitting of the rind, particularly as the cheeses may take up to 6 months or more to reach full maturity.

The finished cheese is expected to have a moisture content of below 42%, and the FDM should be not less than 48%. The dry salting process results in a cheese with 3–4% salt, but this level does not detract from the characteristic and fairly strong flavour of the product. At one time, the flavour was preserved by coating the finished cheese with locally-made tallow pigmented in various ways to give the cheese a red–brown appearance, but aluminium foil has now become the convenient alternative. A rigid outer container is another essential, because the retail cheese – up to 30 cm in diameter, 20 cm high and 12 kg in weight – requires secure conditions for transport and storage.

Mycella

This Danish cheese has a superifical resemblance to Danablu, but the cut surface is usually more creamy yellow than white, and the degree of mould growth may be more intense (see Figure 6.19). As with many of these blue-veined cheeses, the high contents of moisture (48% maximum) and fat (50% minimum FDM) place the products on the borderline between sliceable and spreadable. The flavour of Mycella is usually classed as aromatic rather than biting or tangy, and whether the larger size of Mycella (5 kg) *vis-à-vis* Danablu (3 kg) influences the manner of ripening is not clear. The exact strain of *P. roqueforti* employed for Mycella could be important, as may specific facets of the process, but whatever the precise reasons the products from Denmark are regarded as quite distinct varieties.

Figure 6.18 Gorgonzola, a soft, creamy cheese that is internally mould-ripened; produced widely in Italy from cow's milk. (Courtesy of the Ministero dell'Agricoltura e delle Foreste, Rome.)

Figure 6.19 Mycella, similar to Danablu but a larger cheese; usually the colour of the curd is noticeably yellowish-white. (Courtesy of the Danish Dairy Board, Aarhus, Denmark.)

Roquefort

This sheep's milk cheese is named after the town of Roquefort-sur-Soulzon in south-eastern France and, according to Davis (1976), it is possible that the cheese was first produced around 2000 years ago. Certainly a cheese closely resembling the modern variety was available during the 15th century, and Roquefort is now a protected name with almost world-wide recognition.

The cheese is made from sheep's milk obtained locally, and a local dairy may receive over 15 000 litres a day during peak times – usually during the spring. The raw milk is then subjected to a simple test of microbial quality, and divided into vats of around 1000 litres. After warming to 30 °C, enough rennet is added to bring about coagulation in approximately 2 hours and at this point the curd is cut into cubes. The free whey is drained from the vats, and the soft curd is transferred carefully into perforated metal moulds about 25 cm in diameter and 13 cm high, each holding enough curd to give a finished cheese of some 2.5 kg. As the curd is filled into the moulds, spores of *Penicillium roqueforti* are sprinkled between the layers. The filled moulds are allowed to drain naturally for 4–5 days and loss of whey is encouraged by turning the cheeses several times during this period.

Traditionally, this mass of spores was obtained by inoculating loaves of bread with the fungus and then allowing the mould to colonize the bread totally over a period of 5–6 weeks. The loaf was dried and ground, and the resultant powder sieved and stored for future use.

Removal from the mould is followed by dry salting over a period of one week, giving a salt content of around 4.0% in the finished cheese. To encourage veining, the salted cheese is spiked with some 60 long, steel needles, allowing sufficient oxygen to reach the centre of the cheese and stimulate sporulation of the *Penicillium*. It is only the spores of the fungus that are blue–green; limited oxygen would inhibit spore formation and hence there would be no colour. The salted, spiked cheeses are then taken to the labyrinth of caves – both natural and artificial – that honeycomb the limestone hills overlooking the town.

In this unique environment, the cheeses are stored on their edges on wooden shelves, with each cheese separate from its neighbour so that the natural breezes circulating through the caves can reach every cheese. By chance, this ventilation gives an air temperature that varies between 5 °C and 10 °C and maintains a constant humidity of 95%. These conditions are almost ideal for maturation and, provided that each cheese is surface-cleaned every 2–3 weeks to prevent overgrowth by stray moulds or slime-forming bacteria, the ripening process will be complete in 3–5 months. At maturity, each cheese will be surface-dried and wrapped in foil.

The final cheese (Figure 6.20) should be soft in texture, and the blue veins should be in sharp contrast with the white of curd. The flavour is sharp and slightly peppery, but the short-chain fatty acids typical of sheep's milk provide a distinctive edge. The moisture content is usually around 40%, and the FDM between 48% and 50%; as the milk is not standardized, some variation is both normal and acceptable.

Stilton

Blue Stilton is an English cheese that has to be made in specified factories within the counties of Leicestershire, Derbyshire and Nottinghamshire. The name has been registered by the Stilton Cheese Makers Association who further stipulate that no mechanical pressure may be applied during manufacture, the coat must develop naturally, the form shall be cylindrical and the base milk must be full-cream and come from herds in England or Wales. There is a legal requirement also that the FDM content must be above 48% and the moisture content must not exceed 42%.

Although raw milk was the traditional starting

Figure 6.20 Roquefort, one of the classic blue-veined cheeses. (Courtesy of Food and Wine from France, London.)

point, all milk is now pasteurized at 71.6 °C for 15 seconds (Atkinson, 1993) and cooled to 26–32 °C, depending upon season and other factors. Starter cultures are added at a rate of 0.04% or less and, although each cheesemaker has their own preferred blend of organisms, the essential factor is that acid production should be slow during the early stages of the process. There is a tendency also to include flavour producers such as *Lact. lactis* biovar *diacetylactis* and *Leuconostoc* spp. in the blend. Spores of *Penicillium roqueforti* are added to the milk at this time, and again different factories may select a strain of the mould that best suits their product.

After a lengthy period of ripening (perhaps 1–2 hours), rennet is added. Because the level of starter culture, and consequently developed acidity, is low, up to 1.5 litres of rennet may be needed for each 4500 litres of milk. Animal rennet is normally used; a microbial coagulant is employed in some factories but the texture of the finished cheese does differ from that of the traditional product. Coagulation may take over an hour, and when the curd is firm enough it is cut into cubes with sides of 1.0–1.5 cm. As the curds and whey gradually separate, slow stirring may be introduced to aid syneresis. The curds are then allowed to settle to the bottom of the vat and the whey is slowly drawn off over the next 12–18 hours (Atkinson, 1993). Cutting of the coalescing curd further assists the drainage. At the end of draining, the acidity of the curd may be up to 1.3% lactic acid and the texture firm and easily broken.

Milling follows to give curd pieces of 1.0–2.5 cm in diameter, and salt at a rate of 2.5% (w/w) is mixed into the curd as it exits from the mill. The salted curd is then placed into open-ended moulds (10–11 kg of curd into a mould of 23–25 cm in diameter) and allowed to drain without pressure. As drainage continues, the cheeses are turned at regular intervals – a process that is critical to ensure both even drainage and migration of the salt throughout the cheese. This drainage and turning stage lasts about 7 days, and conditions throughout are maintained at 26–30 °C and 90% RH to ensure that the curd remains sufficiently warm for activity of the starter bacteria, especially the flavour-producers, to continue.

At the end of this initial maturation stage, the cheeses are removed from the moulds and their surfaces are sealed by wiping them with a flat blade. They are then transferred to a cool room (13–15 °C and 85–90% RH) for the curd to lose heat and the surfaces to dry. Similar conditions are maintained for a further 6–7 weeks to allow a firm coat to develop and the cheese to dry internally, while daily (or at least frequent) turning of the cheese encourages uniformity of moisture and texture. Once a firm coat has developed, the cheeses are pierced by 100–300 long, stainless-steel needles to allow air to penetrate the cheese and cause the *Penicillium* to sporulate and form visible blue veins within the curd. The

Figure 6.21 English Stilton, a cheese for the connoisseur. (Courtesy of The Cheeses of England and Wales Information Bureau, UK.)

channels made by these needles are often quite pronounced even at the time of purchase, but, once air has been allowed access, mould growth and sporulation often follow natural fissures in the cheese. Further piercing may be carried out as the mould develops.

After 2–3 weeks, sufficient fungal development should have taken place for the cheeses to be transferred to a cold store at around 5 °C. This reduction in temperature serves to check further mould growth but allows the proteolytic and lipolytic enzymes released by the fungus to continue the process of maturation and flavour development. At 8–9 weeks from production, the cheese should appear like the example shown in Figure 6.21, with a 'flaky' texture and a clean, mild flavour. A stronger flavour does develop with time but, because the impact of the mould is so variable, each individual cheese has to be graded prior to sale.

7 Cheeses surface-ripened with a mixed microflora

There are a number of well-known cheeses which have a distinctive flavour that originates from the activity of a surface microflora, dominated by one specific bacterium, *Brevibacterium linens*. The initial microflora of such varieties may not differ greatly from that of many high moisture cheeses like Brie and Camembert, in that within two or three days of manufacture a range of salt-tolerant, aerobic yeasts, such as *Kluyveromyces*, *Debaryomyces*, *Saccharomyces*, *Candida*, *Pichia*, *Hansenula* and *Rhodotorula* (Lenoir and Tourneur, 1993), along perhaps with the mould, *Geotrichum candidum*, become established on the surface of the cheese. The total figure for such yeasts may exceed one million per gram of cheese by the end of the first week, and their varied activities, such as alcohol production, utilization of lactic acid and release of proteolytic and lipolytic enzymes, both contribute to flavour and encourage the development of other micro-organisms.

However, while the yeasts are overgrown by penicillia in mould-ripened varieties, the absence of surface competition in cheeses not inoculated with *Penicillium* allows *Bre. linens* and other bacteria to become the dominant micro-organisms. These surface bacteria are located in two taxonomic groups, the micrococci and the corynebacteria. The first group is often represented by *Micrococcus luteus*, *M. roseus* and *M. varians*, together with a number of members of the genus *Staphylococcus*, including *Staph. caseolyticus*, *Staph. saprophyticus* and *Staph. xylosus*. The major representative from the corynebacteria is *Bre. linens*, but the genera *Arthrobacter* and *Caseobacter* may be present as well.

The delay in attaining dominance is a reflection of the fact that neither *Bre. linens* nor the other species are acid-tolerant, so that they can grow only when the yeasts and other organisms have utilized the lactic acid in the surface curd and brought about a rise in pH. Once this occurs, an extremely rapid reddish-brown growth of *Bre. linens* becomes visible, and the surface of the cheese may become noticeably 'slimy'; it is for this reason that this group of cheeses are often called 'smear-coated'. The organisms involved, along with *Bre. linens*, are markedly proteolytic during their growth; hence, apart from contributing to the flavour of the end-product, softening of the curd may occur. It is important also that the levels of this surface bacterial flora may reach 10×10^{10} during early maturation, so that its lipolytic and other enzymatic actions can be extremely important in flavour production. The extent to which *Bre. linens* and/or the other species influence the characteristics of the final cheese depends largely upon the opportunity that the organisms have to develop. According to Chapman and Sharpe (1990), the contribution of the surface microflora in general, and *Bre. linens* in particular, is governed by:

- the moisture content of the curd, so that comparatively dry varieties like Monterey show neither a high degree of surface-ripening nor associated flavour development;

- the surface area of the individual cheese that is exposed to the atmosphere during maturation – piling cheeses upon one another tends to limit microbial activity to the sides of the cheese;
- the temperature of the ripening room, and the length of time for which the cheeses are held; and
- any curtailment of the growth of *Bre. linens* by wiping the surface of the cheeses with brine or allowing the rind to become dry.

One or more of these factors can be manipulated to give cheeses that range from mild and sliceable through to strong and spreadable, and any devotee of Limburger will testify that flavour development can be to an extremely high level.

Figure 7.1 Brick, a cheese that originated in North America; the traditional product has its flavour enriched by a surface microflora. (Courtesy of the Wisconsin Milk Marketing Board.)

Brick cheese

Brick is one of the few cheeses to have been developed in the USA, and it provides an excellent example of a cheese where the growth of *Bre. linens* is essential for the final characteristics of the product, but is carefully controlled to avoid a harsh flavour or excessive loss of texture. Consequently, the body of the cheese is softer than Cheddar but still elastic and it slices cleanly, whilst the flavour is mildly aromatic rather than sharp. The name is alleged to have been derived from the brick-like shape of typical cheese (Figure 7.1), which may be some $230 \times 130 \times 140$ cm.

It is manufactured from full-cream cow's milk which is pasteurized and then cooled to 30–32 °C, prior to the addition of a low level of a mesophilic starter culture, and perhaps annatto to modify the colour of the final cheese. In spite of the low level of acid generated, a standard amount of rennet is required to produce a firm coagulum in 30–40 minutes, at which point it is cut into small pieces (<1.0 cm sides) and gently stirred. The temperature of the vat contents is then gradually raised to the scalding temperature – as high as 45 °C is reported to be used in some factories (Davis, 1976) – and the curd is allowed to settle. A portion of the whey is then removed

and replaced with water heated to the temperature of the vat and, in general, the greater the volume of water, the less acid will be the cheese; as more water is added, so lactose is leached out of the curd and acid development curtailed.

The washed curd is then ladled or run into moulds, and a metal or plastic 'plate' is placed on top along with a weight (about 2 kg). Traditionally, this weight might have been a house brick, and some people associate the name of the cheese with this practice. Turning facilitates drainage, and the cheeses are usually left overnight in a warm room (23–25 °C). Next morning the cheeses are dry salted or immersed in strong brine, and a few days later – depending upon the system in use – the cheeses are wiped clean and placed in a store at 14–15 °C and a humidity in excess of 90%.

Under these conditions, yeasts grow actively over the surface during the first two days (Lenoir and Tourneur, 1993), followed by micrococci and, to a lesser extent, *Bre. linens*. The process is assisted in many cases by wiping the cheeses with a brine-soaked cloth. After around 2 weeks, the metabolism of the bacterium will have achieved its purpose and its activity is halted by transferring the cheeses to a drying room. Dipping in wax or wrapping in film completes this initial stage, but further maturation may follow for 2–3 months at 4–10 °C.

A typical analysis might be:

Protein	20–22%
Fat	30%
Fat-in-dry-matter	50% (min.)
Moisture	40–42% (max. 44%)
Salt	1.8–2.5%

On cutting, numerous openings of various shapes are clearly visible.

Butterkäse

This soft cheese is broadly similar to the Italian cheese, Bel Paese, and in both cases a short period of maturation under the influence of a smear coat is essential.

Full-cream cow's milk of good quality is the basic ingredient, and it is normally pasteurized to ensure freedom from bacteria that could give rise to off-flavours. A thermophilic starter including, for example, *Str. salivarius* subsp. *thermophilus* is essential to withstand the temperature of rennet-ing – often as high as 40 °C. At this high temperature, the quantity of rennet can be adjusted to give a firm gel in 10–15 minutes, and the coagulum is then cut into coarse cubes with sides of around 2.5 cm. After a period of stirring at 40 °C to encourage acid development, the curds are moulded and drained.

When sufficiently firm, the cheeses may be brined overnight to achieve a well-defined rind, prior to storage at 5–10 °C for 2–3 weeks. During this time, the smear coat develops to give the cheese its characteristic appearance (Figure 7.2), and occasional washing in brine may be practised in order to ensure an even spread of the bacteria. The surfaces of the retail cheese are sometimes painted with annatto to give a more even yellow colouration. The finished cheeses are wrapped in foil, or occasionally waxed, ready for retail sale. The moisture of the finished cheese can be up to 50%, and the FDM will be standardized in the region of 45–48%. The limited develop-

Figure 7.2 Butterkäse, a butter-like soft cheese that closely resembles the Italian Bel Paese cheese. (Courtesy of the CMA, Bonn, Germany.)

ment of the smear coat ensures that the flavour never becomes too strong, while the high fat and moisture contents endow the product with a pleasing, spreadable texture.

Epoisse and related cheeses

Careful control of the ripening conditions provides cheeses with just the correct degree of maturation, and the effect of more prolonged activity by *Bre. linens* can be seen with the typical example of Epoisse (Figure 7.3). This French cheese has the same soft texture of many smear-coated cheeses, but the golden coat gives it a most distinctive appeal.

Figure 7.3 Epoisse, a soft cheese originating from Burgundy; flavouring with herbs is a common practice. (Courtesy of CIDIL, Paris.)

Figure 7.4 Langres, a soft cow's milk cheese from the Champagne region of France; the attractive appearance results from frequent washing during maturation. (Courtesy of CIDIL, Paris.)

Figure 7.5 Kernhem, a soft cheese that is surface-ripened with bacteria – a process that gives the cheese its pungent aroma and strong flavour as well as an orange colour to the rind. It is sometimes referred to as a 'meshanger' cheese, which means a cheese that 'sticks to the knife when cut'. (Courtesy of the Dutch Dairy Bureau, Leatherhead, Surrey, UK.)

Careful cleaning of the rind during maturation is essential to maintain the appearance of such cheeses, a point clearly evident in a closely-related French cheese – Langres (Figure 7.4). This cheese, which takes its name from a village associated with its origins, is tall and cylindrical with a delightful golden rind and a flavour and aroma characteristic of the genre.

Other countries in Europe have developed similar cheeses, and a typical commercial example is Kernhem (Figure 7.5). This soft (60% FDM), smear-coated cheese from Holland weighs around 1.8 kg, and has a strong flavour and pungent aroma. Even though subjected to supermarket-friendly packaging, the bright orange rind remains an attractive feature.

Esrom

This flat, rectangular cheese (1–2 kg) from Denmark has a semi-soft texture, and its sliceable character is enhanced by the use of a high scald temperature during manufacture. The numerous small, irregular openings (Figure 7.6) are somewhat unusual for a smear-coated cheese, and reflect the fact that the cheeses are held for a while at 15–16 °C after brining. This holding time allows gas production by heterofermentative lactococci and/or leuconostocs from the starter culture to emphasize any mechanical fissures that may arise at the moulding stage.

The flavour is distinctly aromatic, a feature that may be encouraged by the addition of caraway seeds or other herbs. The minimum FDM content is set at 45% or 60%, depending on the type, and the maximum respective moisture values are 49% or 41%.

Figure 7.6 Esrom, a flat, rectangular cheese with a thin, soft rind or a coat of cheese wax; it has a semi-soft but sliceable texture and numerous small, irregular holes. (Courtesy of the Danish Dairy Board, Aarhus, Denmark.)

Havarti

This Danish cheese shares some of the features of Esrom, but the flavour tends to be rather milder and the irregular openings more marked (Figure 7.7). Greater variations in fat content are permitted also, so that while some types have an FDM content as low as 30% (minimum) and a moisture level of 53% (maximum), the relevant figures for other types may be 60% (minimum) and 39% (maximum).

All are made from standardized cow's milk, which is pasteurized and cooled to around 30 °C prior to the addition of mesophilic starter culture, sodium nitrate and standard rennet to give a firm gel in 30–40 minutes. Once cut into small pieces, with sides of approximately 1 cm, the curd is stirred for a short period and allowed to settle. Around 30% of the whey is then drawn off and replaced by warm water to bring the temperature of the vat to 35–37 °C. A low level of salt (perhaps in the region of 0.2%) is added at this point, and the curd/whey mass is stirred for a further 15–20 minutes. After the whey has been run off, the curd is put into cylindrical or, more usually, loaf-shaped moulds. Frequent turning encourages even drainage of the whey.

According to Davis (1976), further development of the starter culture is sought by covering the cheeses with warm water (18 °C) overnight, prior to immersion in strong brine (12 °C) for 2 days. A relatively dry surface is achieved by holding the cheeses at room temperature for 1–2 days, after which time they are transferred to a maturation room at 15–17 °C (>90% RH) for up to 5 weeks. During this time, the yeast and bacterial flora develop, and the cheese acquires a distinctive and aromatic flavour; time, temperature and moisture content determine the final intensity of the flavour.

Once this initial maturation is complete, the rinds of the cheeses are dried and the finished cheeses are wrapped in film. Variants flavoured with herbs and spices are available on the market, as are log-shaped cheeses covered in red film to attract an impulse purchase in the local supermarket.

Figure 7.7 Havarti, a flat, cylindrical or loaf-shaped cheese with numerous small, irregular holes and a thin, soft rind or a coat of cheese wax. Although the cheese is sliceable, it has a semi-soft, supple texture. (Courtesy of the Danish Dairy Board, Aarhus, Denmark.)

Italico

Italico is a cheese from the same family as Bel Paese. Originally it was just an 'Italian soft cheese' but production is now centred on Lombardy, and a standard system of manufacture has resulted in a product of high and consistent quality.

Full-cream cow's milk with at least 3.5% fat is the base material. Following pasteurization and cooling to 45 °C, the milk is inoculated with a thermophilic starter culture that will normally include *Str. salivarius* subsp. *thermophilus*. The high temperature, together with a level of standard rennet that may reach 50 ml/100 litres of milk, leads to a rapid coagulation of the milk, and the gel is then cut into small cubes (1 cm^3). Gentle stirring allows the curd pieces to become firm and, as the whey is drawn off, the curd mass is pressed against the bottom of the vat. Coarse lumps of curd are then transferred to hoops standing on draining-mats. The dimensions of the hoops are chosen to give a finished cheese of 18–20 cm in diameter, 5–7 cm in depth and weighing 1.8–2.0 kg.

Rapid drainage is encouraged by holding the cheese room at 27–29 °C and by repeated turning, so that by the end of the day the cheeses are firm enough to be immersed in brine overnight. A period of cool ripening (5 °C) allows a light smear-coat to develop, assisted with occasional washings with salt water. After around 3 weeks, the cheeses will be mature enough for immediate consumption, although some foil-wrapped products may be held for 2–3 weeks longer.

The finished cheese has a light brown, smooth rind (Figure 7.8), and the texture is soft, slightly pliable and marked by a few irregular openings of mechnical origin. The colour of the cut surface can range from ivory to pale yellow, and the flavour is mild and creamy.

A typical composition might be:

Protein	21%
Fat	26%
Fat-in-dry-matter	50% (min.)
Moisture	50% (max.)

Figure 7.8 Italico, a soft to medium-hard cheese produced in Lombardy from cow's milk. (Courtesy of the Ministero dell'Agricoltura e delle Foreste, Rome.)

Because of the high moisture content, this cheese is best eaten soon after purchase.

Limburger

Although originally produced near (and marketed in) the town of Limburg in Belgium, this cheese is now manufactured extensively in Germany and North America. In some respects it is the classic smear-coated cheese, in that prolonged maturation leads to not only an intense flavour and aroma, but also considerable softening of the curd.

It is made from pasteurized, full-cream cow's milk to which a mesophilic starter culture and rennet are added at 32–34 °C. About 30 minutes later, the curd is cut and gently scalded to around 37 °C. When the curd pieces have become firm, most of the whey is drained from the vat and, in many factories, is replaced by a dilute salt solution; this intermediate brining has the effect of both firming the curd and reducing its acidity. The curd is then scooped into block-shaped moulds capable of producing a cheese of around

Figure 7.9 Limburger, a soft, rindless cheese that derives its yellowish-brown appearance and much of its distinctive flavour from the surface microflora. (Courtesy of the Wisconsin Milk Marketing Board.)

proteolytic enzymes. The end result is a cheese with an extremely strong flavour and aroma and a soft texture.

A typical analysis might be:

Protein	21–23%
Fat	27–29%
Fat-in-dry-matter	50–53%
Moisture	45–48% (50% max.)
Salt	1.8–3.0%

Low-fat Limburger is also on the market, and other national variations are not unknown. Cheeses similar to Limburger are Romadour and Backsteiner from Germany. Another cheese that is sometimes described as a 'high-salt Limburger' is Weisslacker (Figure 7.10), so named because the rind is often white and shiny. This Bavarian cheese has more mechanical openings than Limburger, and a sharp, piquant flavour. The FDM

1 kg (Figure 7.9). Natural drainage, assisted by frequent turning and sometimes the application of light pressure, produces blocks that are firm enough for removal from the moulds and dry salting or, alternatively, immersion in brine at 10–12 °C for 1–2 days.

Maturation follows in a room at 10–15 °C, and the typical flora of yeasts and *Bre. linens* soon develops – a development encouraged by frequent turning of the cheeses and wiping with a brine-soaked cloth. This manipulation is vital in rapidly spreading the microflora from cheese to cheese, a transfer that ensures even ripening and reduces the risk of unwanted contaminant moulds or bacteria colonizing the exposed surfaces. After some 2–3 weeks, the desired microflora will have developed, and the individual cheeses will be wrapped in foil and transferred to a low-temperature store. A further period of maturation for 6–8 weeks allows full development of the flavour – dominated by short-chain fatty acids and a range of volatile organic compounds – and softening of the texture through the activity of

Figure 7.10 Weisslackerkäse, a strongly-flavoured cheese that is similar to Limburger in character. (Courtesy of the CMA, Bonn, Germany.)

standards vary, with 49% being at the top of the range, and moisture contents of 50–55% give rise to the soft texture; the salt content may be in the region of 6–8%. In some regions, the cheese is subject to prolonged periods of maturation to give a product with an extremely strong flavour, and this variant is usually given the name Bier-käse.

Livarot

Livarot is a soft cheese made from cow's milk in Normandy. Production is centred on certain regions around Calvados and Orne, and the name is derived from a village in the area. Each cheese is some 12–15 cm in diameter and 40–50 cm thick, and the best quality cheeses are sold wrapped in grease-proof paper inside individual wooden boxes.

Traditionally, it is made from an equal parts mixture of full-cream and partially skimmed milk, and ripening depends on the natural microflora. The addition of rennet to the warm milk (30 °C) produces a coagulum in around 80–90 minutes, and the entire mass is transferred to a draining bag held in a metal frame. Much of the free whey will drain out over the next 30 minutes, and the curd is then firm enough to put into moulds. Occasional turning helps to consolidate the curd, and the cheeses are then lightly salted over 2–3 days.

As with Coulommier, some Livarot is sold in this fresh state, but most cheeses are moved to a maturation room and regularly wiped with a brine-soaked cloth to encourage development of an intense smear-coat. Once the rind has developed over a period of 2–3 weeks, the characteristic bands of reed are bound around the cheeses (Figure 7.11) and final maturation takes place in storage rooms at 10 °C. The finished cheese has a smooth, pliable texture and strong aroma and flavour.

Figure 7.11 Livarot, a traditional soft cheese of Normandy with a strong aroma and taste. Bands of reeds are employed by some producers to indicate quality – the more bands, the stronger the cheese. (Courtesy of CIDIL, Paris.)

A typical analysis might be:

Protein	26%
Fat	15%
Fat-in-dry-matter	40% (min.)
Moisture	52%
Salt	2.8–3.0%

Although still farm-produced on a limited scale, most Livarot is manufactured in small commercial dairies with a total output of 1000–1200 tonnes per annum.

Mont d'Or

This cheese takes its name from Mont d'Or, near Lyon, and is reputed to have originated several hundred years ago. Although traditionally made from goat's milk, it is now manufactured from cow's milk, and the caprine material is added only as a minor component.

Full-cream or partially skimmed cow's milk, together with a portion of goat's milk, is mixed with rennet. At 32–37 °C, slow coagulation takes place, with the exact time being dependent upon the acid-producing capacity of the natural microflora. When set, the curd may be ladled directly into hoops or cut first; either way, natural drainage on straw or nylon mats is the first stage of consolidation. Turning assists with whey loss, as does the application of light pressure on each cheese. Once it is firm, the cheese is salted on the upper surface. Removal from the mould and wrapping in a decorative, natural band (Figure 7.12) produces a cheese that is sold for immediate consumption. However, a short maturation of 1–2 weeks may be practised, but great care must be taken to ensure that the exposed surfaces remain free from moulds.

A typical composition might be:

Protein	20%
Fat	30%
Fat-in-dry-matter	50–52%
Moisture	42–44%
Salt	2.0%

Figure 7.12 Mont d'Or, a soft, spreadable cheese made from goat's milk or a mixture of cow's and goat's milk in the Lyonnais region of France. (Courtesy of CIDIL, Paris.)

Munster

According to Lenoir and Tourneur (1993), this cheese was produced originally on farms in Alsace and in the vicinity of Munster in Germany, but nowadays large commercial dairies in both Europe and North America dominate the scene. Typically, Muster is a cylindrical cheese (>450 g) with a diameter of 14–20 cm and a height of 3–8 cm (Figure 7.13), but the Little Munster has proved popular with supermarkets (Figure 7.14).

In factory production, full-cream milk at 32–35 °C is inoculated with a mesophilic culture, and the combined action of this culture and standard rennet (30 ml/100 litres of milk) produces a gel in around 40 minutes. After being cut in coarse cubes with sides of around 2.0 cm, the curds are stirred for 45–50 minutes, with occasional rest periods and removal of some of the whey. Moulding is often in large multi-moulds, followed by drainage under gravity assisted by inversion of the moulds on a regular basis. A final rest period overnight consolidates the curd sufficiently for the cheeses to be removed from the moulds and dry salted.

Holding at room temperature for 1–2 days allows the rinds of the cheeses to dry, and they are then placed in a store at 15–16 °C. During this early stage of maturation, the cheeses are

Figure 7.13 Munster, a semi-hard, smear-coated cheese from the Alsace region of France. (Courtesy of the CMA, Bonn, Germany.)

Figure 7.14 Even traditional farmhouse varieties like Munster are now packaged with the supermarket or tourist trade in mind. (Courtesy of Texel, Epernon.)

wiped with warm water every 2 days and about 2 weeks later the growth of *Bre. linens* will have transformed the rind to light orange. A final holding time at 18–20 °C completes the maturation, and the rinds are maintained in a clean state by wiping at least twice a week. The total maturation time, as laid down in the regulations, is a minimum of 3 weeks for Munster and 2 weeks for Little Munster. Observance of this regulation is often achieved by producers delivering 'white' cheeses, i.e. immediately after salting, to specialist companies who complete the maturation process.

Around 8000–9000 tonnes of true Munster are produced annually. Typical FDM values range from 45% to 50%, with moisture levels not exceeding 56%.

Pont l'Evêque

This cheese, named after a town in Calvados, Normandy, was produced originally in the 12th century under the name 'angelot'. Some 500 years later it was given what has become the official name, and today 3000–4000 tonnes are manufactured each year – usually in commercial dairies. It is a square cheese with sides of 10–11 cm and a height of 3–4 cm, although a larger type is produced with sides up 21 cm. It is a soft cheese, and because the mould, *Geotrichum candidum*, often colonizes the surface of the cheese ahead of *Bre. linens*, some authorities tend to speak of it as 'mould-ripened'. However, as with Livarot, the bacterial smear-coat tends to dominate the rind at the end of maturation – hence the light orange colour evident in Figure 7.15. It is not unreasonable to classify the variety as different from Brie or Camembert.

It is manufactured from pasteurized milk with a fat content adjusted to around 3.7%. On cooling to 32–34 °C, a low level of mesophilic starter may be added along with enough rennet to establish the coagulum in 30–40 minutes. Raw milk is still employed occasionally, but the prolonged setting time can prove inconvenient. The gel is then cut into coarse pieces and the curd transferred to a cloth bag or similar draining device for initial drainage. Once firm enough, the curd is packed into perforated metal moulds standing on rush mats. Frequent turning encourages the formation of a firm cheese. After 2–3 days, the finished shape has been achieved, with

Figure 7.15 Pont l'Evêque, one of the traditional cheeses of Normandy, with a recorded history of several hundred years. (Courtesy of CIDIL, Paris.)

The body of the cheese remains firm and pliable rather than spreadable. The flavour and aroma, which can be quite strong, tend to depend on the length of maturation, but the colour of the interior is often adjusted with annatto to a definite yellow hue.

This colour provides a marked contrast with the German cheese Steinbuscher (Figure 7.16), which is somewhat comparable to Pont l'Evêque but milder in flavour. Steinbuscher is also a much larger cheese – up to 1 kg in weight compared with 250–260 g for Pont l'Evêque, and the German regulations set the FDM level at 50%. The pale surface provides a clear indication that maturation has been closely controlled to prevent much bacterial activity occurring subsequent to the growth of *Geotrichum*. This restriction is achieved by drying the surface of the cheeses at an early stage.

Port du Salut

The use of the name Port du Salut is closely restricted, as indeed is the name Port Salut, so that most cheeses of this type tend to be marketed under the name St Paulin. However, to the average consumer, all the cheeses are seen as semi-hard varieties that are compact and elastic in texture, and with a flavour not unlike Gouda; the aroma tends to confirm that a surface micro-flora has been active, as does the colour of the rind (Figure 7.17).

In general, the cheese is produced from pasteurized, full-cream cow's milk – although some factories do manufacture reduced fat variants – to which will be added a mesophilic starter culture and perhaps annatto. After ripening for 25–30 minutes, standard rennet is added with the aim of achieving coagulation in 30–40 minutes at 30–35 °C. The gel is then cut into quite fine pieces of 5–8 mm, followed by stirring and, in some factories, scalding to 40 °C. Once the curds have settled, some of the whey is removed and the curds are stirred vigorously before being transferred to cloth-lined, perforated metal moulds. For St Paulin, some makers replace the extracted whey with brine to give the finished

Figure 7.16 Steinbuscher, a semi-soft cheese with a pleasant flavour and flexible consistency. (Courtesy of the CMA, Bonn, Germany.)

the surfaces scored by the impression of the mats. Dry salting completes the initial processing, and once the surfaces are dry the cheeses are moved to a maturation room at 12–13 °C and a relative humidity of 80–85%. After 2–3 weeks, the cheeses are covered by an even growth of the white mould – encouraged by daily turning – but washing with salty water is used both to inhibit excessive mould growth and to encourage the development of yeasts and bacteria. Consequently, over the 4–6 weeks of maturation, the rinds gradually acquire the characteristics of a smear-coat, and it is this appearance that is most evident to the consumer.

A typical composition might be:

Protein	18–22%
Fat	25–28%
Fat-in-dry-matter	45% (min.)
Moisture	45–50%
Salt	2.0%

Figure 7.17 Port du Salut, originally made in the Monastery of Notre Dame de Port du Salut in Brittany – only cheeses from this Monastery can carry the name. Products manufactured elsewhere are called simply Port Salut. (Courtesy of Texel, Epernon.)

cheese a milder flavour. The filled moulds (18–25 cm in diameter and 70–80 cm deep) are arranged on draining tables, and weighted disks may be placed on each cheese to compress the curd. Regular turning and occasional changes of cloth produce firm cheeses after a few hours, and they are then removed from the moulds and surface-dried. Dry salting, followed by immersion in a brine bath for up to 24 hours, completes the processing stage and the cheeses are ready for maturation.

The cheese stores vary in temperature from 12 °C to 18 °C, according to factory practice, and the RH is adjusted between quite low

and 90%. At regular intervals, the cheeses will be washed with salt water to produce a clean rind, but drying the rind afterwards tends to control the development of the smear-coat. A typical period for maturation might be 6–8 weeks and the cheeses are then ready for wrapping, with or without being cut into retail portions. The FDM value is usually set at 40% for Port Salut and 42% for St Paulin, and the moisture content of St Paulin must not exceed 56% (Davis, 1976).

Reblochon

Although some of the most characteristic cheese is still made on Alpine farms in Haute-Savoie, this output only accounts for around 20% of the 10 000 tonnes produced each year.

Fresh cow's milk is the preferred raw material, and this milk is coagulated in small vats by the action of rennet. The gel is then cut into pieces of about 0.5 cm in diameter, and the curd/whey mass is lightly scalded to 35 °C. After settling, the curd is tipped into cloth-lined moulds around 15 cm in diameter and 5 cm deep, and manually pressed into shape; the addition of a disk and weight (2 kg) completes the shaping process. Over the next 12 hours or so, the cheeses are turned a number of times before being removed from the moulds and dry salted. Storage for 5–6 weeks at 15 °C and 90% RH allows the surface yeasts and bacteria to develop, and regular washing with weak brine prevents excessive development.

The finished cheese, which should weigh 450–500 g, has a yellow/orange rind that is often masked to some degree by a white crust of salt and micro-organisms (Figure 7.18). The minimum FDM figure is 45%, and the moisture content must not exceed 55%; as a consequence, the texture tends towards soft. However, whilst some cheeses like Limburger acquire their soft texture through the action of proteolytic enzymes of microbial origin, Reblochon and a somewhat similar variety, Ridder, derive their character from formulation.

The Norwegian cheese, Ridder, has an FDM value of 60–65%, and this feature, along with a

Figure 7.18 Reblochon, a characteristic cheese from the Haute-Savoie and still essentially of farmhouse origin. (Courtesy of CIDIL, Paris.)

moisture content of at least 40%, gives a semi-soft product that only just remains in the 'slice-able' category. The finished cheese, some 20 cm in diameter and 4.5–5.5 cm in thickness (Figure 7.19) weighs around 1.5 kg, and the pale yellow/orange rind testifies to the presence of *Bre. linens*. The flavour and aroma confirm the presence during maturation of an active microflora, but the extent of such activity is tightly controlled.

Another cheese that is like St Paulin, at least superficially, is St Nectaire (Figure 7.20), but it differs in that production is limited to the Auvergne. Around 11 000 tonnes are manufactured each year, and the best quality product is alleged to be made on farms taking cow's milk from the high mountain pastures. The FDM content must be at least 45%, and the moisture level below 48%. The orange/brown rind confirms the activity of *Bre. linens*, but a variable appearance is not uncommon; on farms, the cheeses are often matured on straw, and hence the microfloras picked-up by the cheeses are not always consistent.

Figure 7.19 Ridder, a cheese of comparatively modern origin, with a characteristic flavour and soft texture. It is surface-ripened with bacteria for around six weeks. (Courtesy of the Norwegian Dairies Association.)

Figure 7.20 St Nectair, a mild-flavoured, semi-hard cheese produced in the Auvergne. (Courtesy of CIDIL, Paris.)

Serra da Estrêla

This cheese is perhaps one of the best known from Portugal, and it is manufactured from sheep's milk, or occasionally from mixtures of milks, along the mountain range of the same name; most production is located within the Serra da Estrêla National Park. There are two main types of Serra da Estrêla cheese: a fresh, soft-textured type which is matured for 30–40 days, and a semi-hard form that is held in store for at least 6 months. The flavour is also defined by the length of maturation, but the fresh cheese has a pleasant and slightly acid flavour, and a marked aroma. The interior is ivory to pale yellow, but with few visible gas holes.

Traditionally, it is manufactured by straining the uncooled, raw milk into any convenient vat (10–20 litres), and then standing the vessel by an open fire in order to achieve the correct tempera-ture of 27–29 °C. Coagulation is peformed by adding to the milk the juice extracted from a thistle, *Cynara cardunculus*. The uncontrolled nature of this extract means that coagulation may take several hours but, once formed, the curd is ladled into open moulds. Pressing by hand expels the excess whey and also forms the curd into the shape of the finished cheese, i.e. up to 25 cm in diameter and 5 cm deep. Once the cheese is firm enough to be removed from its mould, it is dry salted and wrapped in cloth. During the subse-quent maturation period of 1–2 months, the cheeses may be washed with whey and salted again. The nature of the developing microflora changes, in all probability, from farm to farm along with the environmental conditions (Figure 7.21), but government regulations have been introduced with a view to establishing the identity of the cheese.

(b)

(a)

(c)

Figure 7.21 Serra da Estrêla, a farmhouse cheese. (a) The hand-moulded cheese has drained sufficiently to be removed from its mould in preparation for dry salting. (b) The day following salting, the cheeses are placed in cupboards where they are protected from draughts and kept humid; during the main season for cheesemaking (December to April) the average temperature in the storage area or cellar will be 6–8 °C. (c) Serra da Estrêla is sold either as soft cheese after 1–2 months, or as a semi-hard form (Serra velho) after ripening for 6–12 months. (Courtesy of Maria Sousa.)

Taleggio

Taleggio is a soft, smear-coated cheese that was first produced in the Taleggio Valley in Lombardy around 1920, and today around 13 000 tonnes are manufactured each year. It is made from full-cream cow's milk to which a thermophilic starter culture and rennet are added. At 38 °C, coagulation takes place in around 30 minutes; the curd is then cut into coarse pieces and the whey/curd mass is gently stirred. Once the curd has become firm, it is gathered into cloths for initial draining. Moulding gives the characteristic square shape (Figure 7.22) and about 24 hours later the cheeses are dry salted or immersed in brine; the form of the surface imprint is left to the discretion of the producer.

Maturation for 6–8 weeks at 3–4 °C allows the surface microflora to grow and generate a mild, slightly aromatic flavour; the texture is close with just a few small eyes. The composition is laid down by an official Consortium for Taleggio.

A typical cheese will contain:

Protein	19–21%
Fat	27–30%
Fat-in-dry-matter	51–53%
Moisture	48–50%
Salt	1.1–1.3%

Tetilla

Only around 200 tonnes of this cheese are produced annually, and it is the unusual shape (Figure 7.23) that tends to be its main 'claim to fame'.

It is made from raw cow's milk which is held at 20–25 °C and coagulated with standard rennet over a period of 60–90 minutes. The curd is then cut with a knife, and the pieces are allowed to become firm over the next hour. Once the whey has drained out, the curd is pressed by hand and transferred to specially shaped, earthenware moulds. Manual pressure ensures that the curd takes on the desired shape. Once firm, the finished cheese is removed and surface salted.

Figure 7.22 Taleggio, a soft high-fat cheese with origins in Lombardy. (Courtesy of the Ministero dell'Agricoltura e delle Foreste, Rome.)

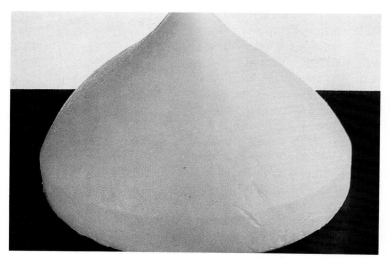

Figure 7.23 Tetilla, produced in Galicia in northern Spain – the unique shape is the result of hand-moulding. (Courtesy of the Ministerio de Agricultura Pesca y Alimentacion, Spain.)

Maturation at 10–15 °C allows the smear-coat to develop, but a comparatively low RH (70–75%) restricts the activities of the essential yeasts and bacteria.

Although not subject to standardization, a typical analysis might be:

Protein	22%
Fat	33%
Fat-in-dry-matter	54%
Moisture	39%

Tilsit

The town of Tilsit in Germany has given its name to this important variety of cheese. While the flavour has been described as 'mild Limburger', the texture is much firmer and the mechanical openings are more pronounced (Figure 7.24); the colour of the cut surface varies with fat content from white to bright yellow.

It is manufactured from full-cream or partially skimmed cow's milk, and pasteurization followed by cooling to 32–34 °C is now the standard preliminary treatment. A mesophilic starter culture is added to provide the essential acidity,

followed by standard rennet (25–30 ml/100 litres of milk) and often calcium chloride; as usual, the latter assists with the formation of a firm coagulum. About 45 minutes later, the gel is cut into

Figure 7.24 Tilsit (Tilsiter), a smear-coated cheese that was first produced near the town of Tilsit in Prussia. (Courtesy of CMA (UK), London.)

small pieces and gently stirred for 30–40 minutes with a modest rise in temperature to cause some shrinkage of the curd. As the whey is drained, so the curds are tipped into moulds standing on an appropriate draining table. The temperature of the room is held at 32 °C to encourage the continued metabolism of the starter organisms, and frequent inversion of the moulds promotes drainage of whey, as does the absence of mechanical compaction of the curd. Over the next 2 days the cheeses are salted, usually by immersion in brine at 15 °C, and then air-dried. Maturation at 15 °C (>90% RH) for at least one month allows full development of the smear-coat, a process encouraged by regular washing of the cheeses with brine; however, the rind is rarely allowed to develop the intense orange colour often associated with the growth of *Bre. linens*.

Although originally a wheel-shaped cheese, the block form is more popular with retailers and the same shape allows for easy packaging in the factory. The composition is permitted to vary quite widely, with FDM levels ranging from 49% to 57% according to type, and moisture contents of 45–55%; the salt content is usually around 2.5%.

The popularity of ripening under a smear-coat is demonstrated by another German cheese, Wilstermarsch, which has a mildly acid flavour with just a 'hint of Tilsit'. Mechanical pressure provides Wilstermarsch with a more compact structure (Figure 7.25), and salting of the curds before moulding also assists in the expulsion of whey. A short maturation time compared with Tilsit holds back flavour development, while the attractive film-wraps have raised its profile on the supermarket shelves. FDM values range from a low of 20% to as high as 53% depending on the type, and moisture contents are permitted as high as 55%.

Trappist

This cheese was first manufactured by Trappist monks, but it is now manufactured quite widely throughout Europe. It is a pale yellow cheese with a mild, aromatic flavour, and the texture

Figure 7.25 Wilstermarschkäse, a semi-hard cheese that is not unlike a mild variety of Tilsit; it may be made from full-cream or semi-skimmed milk. (Courtesy of the CMA, Bonn, Germany.)

varies from soft to firm depending on the degree of mechanical pressure applied after moulding; fissures of mechanical origin are usually prominent (Figure 7.26).

Originally, Trappist was made from any milk available or mixtures thereof, but nowadays pasteurized cow's milk has become the usual starting point (Scott, 1986). Annatto is sometimes added along with calcium chloride, and a mesophilic starter culture provides the acidity necessary for the action of the rennet. A ripening temperature of 30–31 °C is usual, with a slight elevation to 34–35 °C to bring a degree of firmness to the cut curd. After being stirred for 40–45 minutes, the curd is allowed to settle and the whey is replaced with clean water at the same temperature. Stirring the curd for 15–20 minutes leaches out enough of the lactose/lactic acid to ensure that the finished cheese retains a mild flavour, at which point the curds are moulded. Whey drainage by gravity rather than mechanical pressure produces a cheese sufficiently firm to immerse in brine at 15 °C for 7–8 hours; in some countries, large cheeses are produced that are pressed in the mould to expel excess whey.

A short maturation period of 2–3 weeks at 16–18 °C is sufficient for a light smear-coat to develop, but for the most part it hardly affects the flavour of the final cheese. Dipping the cheeses in hot wax is a favoured means of preservation for storage and sale, and typical composition might be:

Protein	23%
Fat	26%
Fat-in-dry-matter	48%
Moisture	46%
Salt	1.5–2.5%

Figure 7.26 Trappist cheese, believed to have originated in a Trappist monastery in Croatia but now widely produced in Germany and elsewhere. (Courtesy of the CMA, Bonn, Germany.)

8 Some typical varieties of fresh cheeses

The earliest forms of fermented milk must have arisen when milk was allowed to sour naturally, and some of the end-products may well have been consumed without any drainage of the whey. Such products are still found today, and whether the coagulation is achieved by bacterial and/or enzyme action or the direct addition of acid, the weak gel can be sold as a simple 'fresh cheese'. As no drainage of the whey is desired, the products are often solidified in the retail containers, and are regarded as cheap, short shelf-life items to be eaten with fruit or salads.

Although many people find these 'junket-style' cheeses pleasant, the majority view might well be that a product with more body or flavour has more to recommend it. Consequently, most fresh cheeses have developed beyond this simple stage and, although most varieties are not resistant to spoilage and must be consumed within days of purchase, attention to texture and flavour is much in evidence.

In general, mesophilic starter cultures are employed along with standard rennet to produce the initial coagulum, and acid production by subspecies of *Lac. lactis* is a prime requirement. Whether or not the curd is cut and scalded during manufacture depends upon the variety, but the stages of moulding and whey drainage without pressure tend to be universal. Maturation, if it occurs at all, will be limited to a matter of days, and the cheeses must then be distributed rapidly to the retail markets.

Burgos

This Spanish cheese is made in the region around Burgos, and it is the main soft cheese made from sheep's milk. Traditionally, raw milk was employed, but nowadays the trend is to use pasteurized milk and to extend the volume of production by mixing in a proportion of cow's milk. Coagulation is achieved by the addition of rennet and calcium chloride alone. At 28–32 °C, a gel should be formed in 60–90 minutes; the precise time depends, to some extent, on the natural acidity of the milk. The coagulum is allowed to achieve additional firmness for a further 30–40 minutes, before being cut into pieces with sides of around 2.0 cm; gentle stirring and heating to 38 °C follows. Once the curds have been separated from the whey, they are transferred to earthenware moulds and pressed into shape by hand. The result is a flat, cylindrical cheese (1–2 kg) of around 15 cm diameter and 5–7 cm depth, with markings taken from the walls of the mould (Figure 8.1).

The finished cheeses, which may remain unsalted or, alternatively, be immersed in brine for 10–30 minutes, are stored at 4 °C prior to sale; consumption should take place within 3–5 days of manufacture. Around 10 000 tonnes of this variety are produced annually.

A typical composition might be:

Protein	16%
Fat	24%
Fat-in-dry-matter	52%
Moisture	54%

Villalon cheese is broadly similar to Burgos.

Cottage cheese

This cheese has risen in popularity from being of real interest only to farmers as a means of using

Figure 8.1 Burgos, a soft cheese made in the region around Burgos; it is consumed within a few days of production. (Courtesy of the Ministerio de Agricultura Pesca y Alimentacion, Spain.)

the skimmed milk left over from cream separation, to the status of one of the fastest-growing retail items in the chill cabinet. This success is due mainly to the widespread use of skimmed milk as the raw material for manufacture, so that calorie-conscious consumers are provided with a low-fat cheese that is compatible with many diets. The bland flavour of the product has encouraged its admixture with herbs or even fruit, to give a product with marked versatility.

Although full-fat types are available, most commercial products are based upon skimmed milk which is pasteurized and cooled to the desired temperature of incubation. This value can be as low as 20 °C or as high as 35 °C (Scott, 1986), but 30 °C is probably used most widely. The choice of temperature is determined in part by the process, because manufacturers have two options available – the long-set or the short-set method. In the long-set method, a low level of mesophilic starter culture is used (perhaps 0.5% v/v for a liquid culture) and the intention is that coagulation of the milk should take place over-

night (14–16 hours). By contrast, the short-set method envisages gel formation in 5–6 hours, and both temperature of incubation and inoculation rate (perhaps 5.0% v/v) reflect this intention. In both cases, the addition of a low level of rennet (1–2 ml/450 litres of milk) may be employed to assist coagulation, but destabilization of the casein by lactic acid is the essential element.

Once the gel is firm and the pH has fallen to 4.5–4.6, it is cut into pieces of the dimensions desired by the market – some markets prefer Cottage cheese with small granules, other consummers expect large particles. After a resting period of around 30 minutes for the surfaces of the curd pieces to harden slightly, the temperature of the vat is raised to 50–54 °C over a period of some 2 hours. Gentle stirring is employed to ensure that the curd remains in the form of discrete granules.

When the particles have become firm, the whey is drained out and replaced by clean water at 7–10 °C. This washing removes the residual lactose from the curd to prevent late acidification, and the process may be repeated three times in all, with the last wash in water at 1–2 °C. This latter wash also lowers the temperature of the curd, and final drainage is often delayed for up to an hour in order to achieve this effect. Such prolonged exposure means that the water must be of excellent quality both chemically and microbiologically, for otherwise taints or the growth of slime-forming bacteria can quickly destroy the finished cheese. After the curd has been thoroughly drained, it may be mixed with other ingredients prior to packaging (Figure 8.2) and storage at 1–2 °C for distribution and sale. The finished cheese should be white in appearance (Figure 8.3) and even if the granules are masked by a dressing in the carton, they should become distinct in the mouth to give the cheese a somewhat 'chewy' texture.

The fat content of Cottage cheese (low fat) will be in the range of 2–10%, depending upon the brand.

Figure 8.3 Cottage cheese, typically made with skimmilk; the bland flavour can be easily enhanced with various dressings or other ingredients, e.g. chives. (Courtesy of the Wisconsin Milk Marketing Board.)

A fairly typical analysis (Shaw, 1993) might be:

Protein	14%
Fat	4%
Fat-in-dry-matter	19%
Moisture	79%
Salt	1.0%

Cream cheeses

There is a range of products on the market with names like Fromage frais, Cream cheese and Quarg in which the essential features of manufacture are:

- acid coagulation of skimmed or full-cream milk, perhaps with the assistance of a low level of rennet;
- cutting of the gel and separation of part of the whey by centrifugal action or membrane filtration;
- blending of the curd to give a homogeneous mass;
- packaging into cartons or compressing into blocks, depending on the final total solids of the cheese.

In general, fat contents vary from less than 2%

Figure 8.2 Cottage cheese – especially the low-fat variety – is popular with diet-conscious consumers. (Courtesy of Texel, Epernon.)

Figure 8.4 Mainzerkäse, a soft, sour-milk cheese which is sometimes given a short maturation to increase the level of flavour; the fresh version may be sold under the name Quargel. (Courtesy of the CMA, Bonn, Germany.)

for a skimmed-milk cheese to over 45% for true Cream cheese, with perhaps 22–24% (44–45% FDM) being about average; in the USA, however, the fat content must be 33% or higher, and the cheese is made from a mixture of cream, milk and skim-milk powder (the mix is homogenized prior to fermentation). Moisture contents can be equally variable, with 80% being the likely maximum and 50% about average. Mesophilic starter cultures are employed almost universally to provide the desired acidity.

As a sector of the chilled food market, these types of soft cheese are of increasing importance, and Figure 8.4 shows a typical example of a compressed product. In some cases, brand names like Philadelphia have become synonymous with a type of cheese, but they are all, in effect, unripened soft cheeses. Elsewhere, consumer convenience is employed to assist with marketing, and the products shown in Figure 8.5 are good examples of how the basic cheese can be modified to provide additional market outlets. Natural or fruit-flavoured, 'spoonable' soft cheeses have also

Figure 8.5 Soft cheeses are packaged for the supermarket with a wide variety of condiments. Here, layers of flavouring materials (herbs of Provence, or gherkins and paprika) provide the inducement to purchase. (Courtesy of Texel, Epernon.)

Figure 8.6 The pyramidal shape of this cheese is typical of many goat's milk products from France. (Courtesy of Texel, Epernon.)

become widely popular as snack foods, and another variant is the 'semi-solid' form shown in Figure 8.6. A number of products with this characteristic shape are available, and many are sold in cartons rather than wrapped as single cheeses. The advantage of carton sales is that the total solids of the milk can be raised, prior to coagulation, to that desired in the finished cheese, so that once the carton has been filled, no further handling of the product is necessary.

This latter point can offer advantages to consumers and manufacturers alike, in that one of the problems with soft cheeses is their relative susceptibility to spoilage. The combination of high moisture content and low/medium levels of acidity means that spoilage by bacteria and/or fungi can be appreciable at temperatures above 4 °C. High standards of hygiene during manufacture, together with an effective chill chain for distribution, should avoid major incidents of spoilage, but consumers should be aware of the need for careful handling after purchase. To this end, it is noticeable that some types of cream cheese are now pasteurized after manufacture and 'hot-filled' into the retail containers. Obviously this step modifies, to some degree, the characteristics of the product, but it does offer advantages with respect to safety and anticipated shelf-life.

Figure 8.7 Cremoso, a very soft cheese that is consumed within 2–3 weeks of manufacture. (Courtesy of Dr Carlos Zalazar, Instituto de Lactologia Industrial, Argentina.)

Cremoso

Pasteurized cow's milk, standard rennet and a mesophilic starter culture are used to manufacture this very soft, square cheese (Figure 8.7). The final product is around 7 cm thick, and it may be sold loose with the surface coated in maize oil, or it may be vacuum-packed in plastic bags that are capable of heat-shrinkage; the tight adhesion between the plastic and the cheese lowers the risk of mould growth on the surface of the retail item.

It retains its creamy texture for about 15–20 days at 7–8 °C, but after this time it may become over-ripe and almost liquid as the result of proteolysis; continuing activity of the residual rennet is believed to be responsible for the observed deterioration. It is widely used in Argentina as a topping for pizzas or other dishes, but may be used also as a table cheese.

Crescenza

Stracchino is a generic name applied to many soft cheeses in Italy, and Stracchino Crescenza (Figure 8.8) is a soft, creamy variety produced in Lombardy.

Production is usually a simple process, with low levels of starter culture and rennet being added to full-cream cow's milk at 30–32 °C.

When the milk has coagulated, the gel is cut into coarse segments and ladled into oblong moulds standing on draining-mats; each mould should be large enough to give a finished cheese of the weight desired by the factory, namely from 200 g up to 1.5 kg. After drainage for two days, the cheeses are removed and dry salted. A short holding/ripening time of 10–15 days at 5–7 °C may precede retail sale, and hence an element of surface-ripening by bacteria cannot be ruled out; the FDM content should exceed 50%.

Fior di Latte (Figure 8.9) is another soft, unripened Italian cheese (although its 'stringy' texture is rather similar to Mozzarella) and the distinctive shape adds to its consumer appeal. It is made all over Italy from full-cream cow's milk, and the initial coagulation takes place at 35 °C under the combined influences of starter culture and rennet. Shaping into cheeses of 50–1000 g follows, and the manner in which the curd is transferred often gives the cheese a 'layered' appearance. It has a mild, slightly sour flavour, and the high moisture content confirms that the product is for immediate sale and consumption.

A typical analysis might be:

Protein	17%
Fat	19%
Fat-in-dry-matter	44%
Moisture	57%

Figure 8.8 Stracchino Crescenza, a soft, high-fat cheese produced in Lombardy from cow's milk. (Courtesy of the Ministero dell'Agricoltura e delle Foreste, Rome.)

Figure 8.9 Fior di Latte, a traditional variety dating back to the 15th century, is a soft cheese composed of layers and similar to Mozzarella. (Courtesy of the Ministero dell'Agricoltura e delle Foreste, Rome.)

Galotiri

Galotiri is claimed to be one of the oldest cheeses of Greece, and it is manufactured in limited quantities, often on farms, throughout the country. Sheep's milk is the raw material of choice and, after boiling, it is placed in an earthenware vessel and left to stand for about 24 hours. Salt, at a rate of 3–4%, is then stirred into the milk. After standing for a further 48 hours, with occasional stirring, the milk is poured into an animal-skin bag or wooden barrel.

Further additions of boiled and salted milk are made over the next few days until the vessel is full, when it is left undisturbed for the milk to set and the whey to drain out. Sometimes a small piece of Feta cheese may be added as a 'starter culture' along with a little rennet but, even so, coagulation and drainage may take several weeks. Eventually, a soft, sour-tasting cheese (pH 3.8–4.0) is derived (Figure 8.10).

A typical analysis might be:

Protein	10%
Fat	14%
Moisture	70%
Salt	2.7–3.0%

Figure 8.10 Galotiri, a traditional soft, spreadable cheese made from sheep's and/or goat's milk. (Courtesy of Professor E.M. Anifantakis.)

Despite the high acidity, it is not a cheese that stores well once it has been removed from the skin bag or barrel.

Karish

This soft, skimmed-milk cheese has been made in Egypt for hundreds of years, and Abou-Donia (1991) suggests that a similar cheese was probably being produced at least as early as 3000 BC.

The first stage of manufacture is to produce a sour milk known locally as 'laban rayeb'. At farm level, the inside of an earthenware pot is rubbed with sour milk, and the pot is then dried overnight in an oven. Whether or not the sour milk provides an inoculum of bacteria is not clear but, next day, fresh cow's or buffalo's milk is poured into the pot and left to stand at ambient temperature. Although temperature control is minimal, the milk must not be allowed to become too warm or it will coagulate before the fat has risen to the surface. Normally, 2–3 days are required for separation of the fat and coagulation of the skim-phase, at which point the fat is removed for the production of butter or ghee; the coagulated skim-phase is 'laban rayeb'.

Figure 8.11 Karish, a soft cheese that has been produced and consumed in Egypt for thousands of years. (Courtesy of Professor S.A. Abou-Donia.)

The laban rayeb is then tipped on to a mat made of reeds and, after a short period of draining, the edges of the mat are drawn together to force out more of the whey. The total drainage time may be 2–3 days, and the mass of the curd is then formed by hand and cut into small squares (Figure 8.11). The individual pieces of Karish are dry salted prior to immediate sale or consumption.

A typical chemical composition might be:

Protein	11%
Fat	5.8–6.0%
Moisture	70%
Salt	4.8–5.0%

Local variations are the rule rather than the exception.

Kopanisti

Kopanisti is a traditional Greek cheese made from cow's, goat's or sheep's milk – or mixtures of all three – on the Cyclades Islands (Anifantakis, 1991). It has a soft texture and a salty, piquant flavour which, in good quality batches, has much in common with Roquefort.

Raw milk is employed for manufacture, and at a temperature of 28–30 °C, sufficient rennet is added to give a coagulum in 100–200 minutes. As the only acidity is derived from the natural microflora of the milk, the uncut gel is allowed to stand for 24 hours to become both firm in nature and acidic in reaction. The coagulum is then cut and ladled into bags for draining, a process usually accelerated by placing stone weights on the bags, or piling the bags on top of each other.

After draining, the curd is tipped into a convenient vessel and thoroughly blended to an even consistency along with 5% salt. The salted curd is then placed into basins and left to stand at ambient temperature for around one week. At the end of this period, mould growth is clearly visible on the surface of the cheese, and the basins of curd are mixed manually or mechanically to distribute the fungus throughout the mass of cheese. This overgrowth by moulds and their distribution into the cheese may take place 3–4

Figure 8.12 Kopanisti, a traditional soft cheese made from any available milk; it has a spreadable consistency and an intensely salty, piquant taste. (Courtesy of Professor E.M. Anifantakis.)

times, and it seems likely that the 'peppery', Roquefort-like flavour of good quality Kopanisti is derived, at least in part, from this microbial activity. Indeed, so critical is the activity that it could be argued that Kopanisti should not, aside from its textural features, be placed alongside the other soft cheeses. However, as shown in Figure 8.12, the homogeneous physical texture of Kopanisti does permit a certain degree of tolerance as to its 'classification'.

After final mixing, the cheese is packed into clay vessels (Figure 8.12), covered with a waterproof paper and placed in a cold-store ready for immediate distribution to the market. The lipolytic and proteolytic activities of the mould growth tend to modify the composition of the finished cheese to a considerable degree, and in a manner that varies from batch to batch, but the choice of raw material exerts an even more decisive influence. Consequently, the chemical composition can, unlike the situation with many cheeses, be subject to wide variation.

A typical analysis might be:

Protein	16–17%
Fat	19–20%
Moisture	60%
Salt	3.0%

Queso Blanco

There are many cheeses around the world that are made by the direct acidification of milk with lemon juice, vinegar or other agents, and Ricotta from Italy is perhaps the most widely publicized example. However, many countries with warm climates have developed similar types (Chandan, 1991), and Queso Blanco is a 'catch-all' term for the many types of 'white cheese' produced throughout Latin America.

Some varieties are produced on farms, others in factories, and each country or region tends to give its cheese a local name. In some places, heat and acid give rise to the curd, while other makers

Figure 8.13 Queso Blanco, the popular soft cheese of Latin America, produced in many forms throughout the region. (Courtesy of Alfa Laval, Lund, Sweden.)

have now adopted rennet and ambient temperatures to produce the coagulum. Either way, the end-product is usually a firm, salty cheese with the form shown in Figure 8.13.

A typical farm-based procedure was described by Davis (1976) as involving:

- pouring fresh, warm milk into a convenient vessel;
- coagulating with rennet and cutting into coarse fragments;
- allowing the curds to settle, and stirring salt (around 5%) into the whey;
- leaving the curds and whey to stand overnight;
- draining off the whey, and mixing more salt into the curd;
- packing the curd into square moulds and pressing.

The manual packing operation and the absence of mechanical presses account for the numerous openings in a typical cheese (Figure 8.13), as well as the soft, pliable texture of many batches.

Many such cheeses are eaten within a day or

two of production, and two weeks would be the maximum shelf-life for many types. Nevertheless, these cheeses are a valuable component of the diet.

A typical analysis might be:

Protein	23%
Fat	15%
Fat-in-dry-matter	33%
Moisture	55%
Salt	2.5–3.0%

Torta del Casar

Although the climatic conditions of Spain favour the production of harder varieties, Torta del Casar is a soft cheese made regularly from sheep's milk in the region of Cáceres. Farm production is normal for this variety, and the first stage involves collecting the raw milk and heating it in any convenient vessel to 32–35 °C. As acidification depends on the natural microflora – often a mixture including subspecies of *Lac. lactis*,

Figure 8.14 Torta del Casar, a soft cheese made from sheep's milk in the region of Cáceres. (Courtesy of the Ministerio de Agricultura Pesca y Alimentacion, Spain.)

Leuconostoc spp. and *Lactobacillus plantarum* – coagulation with extracts from the thistle, *Cynara cardunculus*, may take 2 hours or more. The gel is then cut into coarse pieces, perhaps 2–3 cm along the sides, and the curds and whey are left to stand for a prolonged period for further acid development to take place.

After perhaps 10 hours, the curds are ladled into moulds that will give a finished cheese of 25–35 cm in diameter and 4–7 cm in height and between 0.9 and 1.2 kg in weight (Figure 8.14). Natural drainage is sufficient to expel the excess whey, and the finished cheeses are then dry salted and stored at 25 °C for 2–3 weeks. Some superficial growth of yeasts and bacteria may take place during this time but, as the relative humidity is held around 55%, the contribution of any microflora to flavour is unlikely to be dramatic.

Whey cheeses

Along with Anthotiros (described in Chapter 6), Greece has two other famous cheeses: Manouri and Mizithra, which are made either from fresh milk or from whey derived from other cheese-making processes. Anthotiros was cited under the category of mould-ripened cheeses because much of the limited output is matured prior to sale, whereas Manouri is sold in the fresh state; Mizithra can be sold either mature or fresh, but the balance is probably in favour of the fresh state.

Mizithra

According to Anifantakis (1991), around 500 000 tonnes of whey are produced in Greece each year from primary cheesemaking, and hence it is not surprising that the manufacture of whey cheeses has become a major industry.

The whey from sheep's milk is the most widely used starting point. After filtration to remove any curd particles, the clear whey is poured into vats of around 1000 litres capacity. Heating, initially with stirring, follows until the whey has reached a temperature of 88–92 °C and at this point the proteins in the whey are denatured and rise to the surface of the vat. In order to improve the quality of the final cheese and increase yields, full-cream

Figure 8.15 Mizithra, a soft cheese made traditionally from whey and consumed either fresh or, after drying, as a topping for main dishes. (Courtesy of Professor E.M. Anifantakis.)

milk (3–5% v/v) may be added to the vat during the heating stage, along with 1.0–1.5% salt. The vat contents are held at high temperature (without stirring) for 20–30 minutes, and then the protein mass is skimmed off the surface of the liquid and into cloth-lined moulds – cone-shaped with a basal diameter of 12–13 cm and a depth of 16–18 cm.

Drainage is complete in around 5 hours, and the fresh cheeses (Figure 8.15) are placed in a cold store ready for delivery to market next day. If the 'dried' type is required, then the cheeses are dry salted and stored in cool rooms in the manner of Anthotiros (Chapter 6). The chemical composition of this variety depends on the additions made during production, e.g. volume and fat content of the added sheep's milk, so that the figures given in Table 8.1 are prone to wide variations.

Manouri

Originally, Manouri was produced in Macedonia from full-cream goat's milk or from mixtures of sheep's and goat's milks, and its manufacture from whey is a relatively recent innovation. The

Table 8.1 Typical compositions (%) of fresh whey cheeses

	Manouri	Mizithra	Anthotiros
Protein	10	13	10
Fat	37	16	17
Moisture	48	66*	68
Salt	0.8	0.8	–

*On drying, this value will fall to around 39%.

Figure 8.16 Manouri, a soft cheese with a fat content in excess of 30% and a distinctive shape. (Courtesy of Professor E.M. Anifantakis.)

process is much like that for Mizithra, except that the overall heating time may be longer, as will the drainage stage; both changes are necessary to reduce the moisture content of the finished cheese. Cloth bags are employed for drainage, and the dimensions are such that the final cheese will have a diameter of 10–12 cm and a length of 20–30 cm (see Figure 8.16). The high fat content of Manouri means that it must remain chilled during distribution to avoid off-flavours developing and/or leakage of the fat.

Some typical chemical compositions of these whey cheeses (fresh) are given in Table 8.1. The acidities of all the varieties are mild – the typical pH is 6.0 – so that early consumption of the fresh cheese is important; within 2 weeks at 4 °C is the aim, but vacuum packing may provide a modest extension of the storage time.

Ricotta

The name 'Ricotta' is derived from the Italian word *ricottura*, which means 'reheated' – the cheese was made originally in Italy from whey which was heated to precipitate the protein. Wheys from the production of cheeses like Mozzarella or Provolone were the most popular raw materials, and the whey, together perhaps with some full-cream milk, would have been heated to 85–90 °C in a convenient vat. Salt and an acidifying agent like lemon juice or vinegar would have been added to help precipitate the denatured protein. After a short holding time, the protein and entrapped fat would have been skimmed off into chilled, perforated moulds (Figure 8.17). The consistency of the finished cheese was controlled by the duration of drainage, and perhaps the use of weights to apply a light pressure, so that while some cheeses became sliceable, others were best retailed in cartons (Figure 8.18).

However, although this traditional cheese is available, the name 'Ricotta' is nowadays applied to varieties manufactured by the heat-precipitation of the proteins from full-cream or skimmed milk. In Italy, sheep's or cow's milk is the usual starting point, and fresh cheeses from 300 g up to 3 kg (Figure 8.19) are widely avail-

Figure 8.17 Ricotta – after transfer of the curd to the moulds for draining, ice is packed around the containers to cool them. (Courtesy of Professor F.V. Kosikowski.)

Figure 8.18 Ricotta, widely produced in central and southern Italy and traditionally made with whey, but increasingly manufactured with whole milk. (Courtesy of the Wisconsin Milk Marketing Board.)

Figure 8.19 Ricotta, a soft, fresh cheese. (Courtesy of the Ministero dell'Agricoltura e delle Foreste, Rome.)

able in the market; the best quality cheese is reported to be made from sheep's milk. The texture is soft and the flavour slightly sour and, as the analysis suggests, the shelf-life is not prolonged. However, a salted and matured (60 days minimum) variant is produced, and this type would be expected to show more in-store stability.

A typical analysis of Ricotta made from fresh cow's milk might be:

Protein	18%
Fat	8%
Fat-in-dry-matter	30%
Moisture	73%

9 Yoghurt and other fermented milks

Traditionally, yoghurt was a semi-solid product of high acidity that was consumed with bread or some other component of a main meal, but in recent years this usage has been largely overtaken by its conversion into a snack food. This change has been encouraged by a number of factors, but the two principal ones are probably:

- the introduction of fruit and fruit-flavoured yoghurts, which brought yoghurt to a much wider band of consumers; and
- the development of convenient, single-portion cartons, so that yoghurt became available as a dessert for picnics or lunch-time snacks.

Consequently, the market for yoghurt has developed over the last 40 years into a multi-million dollar industry, and one that is still expanding as new 'variations on the theme' are developed and promoted.

Nevertheless, the basic types of fermented milk are still recognizable, and Robinson and Tamime (1991) classified them according to the important components of the microflora (Table 9.1).

Products derived by the bacterial fermentation of the lactose in milk to lactic acid are by far the most popular world-wide, and the most convenient subdivision is on the basis of the dominant microflora.

Mesophilic lactic fermentations

Although leben found in North Africa shares much the same microflora, the products in this section are probably north European in origin. The low optimum growth temperature (20–22 °C) of the important bacteria tends to support this view, and most employ one or more cultures selected from the following:

- *Lactococcus lactis* subsp. *lactis*
- *Lactococcus lactis* subsp. *cremoris*
- *Lactococcus lactis* biovar *diacetylactis*
- *Leuconostoc mesenteroides* subsp. *cremoris*

Table 9.1 Classification of types of fermented milk (Robinson and Tamime (1990))

Type of lactic fermentation	Fermented milks
Yeast	Kefir
	Kumiss
Mould	Villi
Mesophilic	Buttermilk
	Lactofil
	Filmjolk
	Nordic ropy milk
	Taetmjolk
	Ymer
Thermophilic	Yoghurt
	Labneh
	Skyr
	Bulgarian buttermilk
'Therapeutic'	Biogarde*
	Bio-yoghurt
	Acidophilus milk
	Acidophilus drinks
	Cultura*
	BA products
	Yakult

*Trade names.

The role of these cultures is broadly along the lines indicated in Chapter 1, in that whilst the former two are mainly considered as 'acid-producers', the latter organisms are responsible for the butter-like flavour and, on occasions, limited effervescence associated with these mesophilic milks.

Buttermilk

Traditionally, buttermilk was the liquid/skim-milk fraction that remained after the churning of cream into butter, and often this component was allowed to ferment naturally into a mildly acidic drink with a distinctive buttery flavour. Low levels of carbon dioxide provided an additional refreshing note to the product, and many farming communities must have found it an enjoyable addition to their diets.

Nowadays, skim- or semi-skimmed milk provides the normal base. This milk is homogenized and heat-treated at 90–95 °C for up to 10 minutes. After being cooled to 22 °C, the milk is inoculated with a predetermined mixture of 'acid-producers' and 'flavour-enhancers' and incubated until the pH of the milk is around 4.6. The product can then be cooled to 2–4 °C and packaged into cartons of the type employed for normal liquid milk. Cool storage ensures that there is no excessive build-up of carbon dioxide and bulging of the packages but, even so, the shelf-life tends to be fairly restricted.

Having the consistency of ordinary milk, buttermilk tends to be consumed as a 'thirst quencher', rather in the manner of its Middle Eastern counterpart, leben, and hence market demand depends upon the weather and the availability of alternatives, such as lemonade or cola. The addition of flavours and colours has had little impact on retail sales, at least in the USA, but it could have potential as a raw material for further processing. Its use in baked goods has been recommended, while a low-level incorporation of spray-dried buttermilk powder into cheese milk can improve the flavour profile of certain white-brined cheeses (M.S.Y. Haddadin, personal communication).

Ymer

Ymer is a cultured milk product with around 3.5% fat and a protein content elevated to 5–6%, so that the end-product is not unlike a thick, stirred yoghurt. Traditionally, the principle of manufacture was rather similar to that for the production of labneh (described later), in that milk was fermented and the gel so obtained was further treated in order to encourage much of the whey to drain free. The major difference was that while whey drainage during labneh production involved hanging the coagulum in a cloth or animal-skin bag, ymer was produced by simply warming the gel to encourage syneresis/whey loss. If ymer was made from skim-milk, then cream could be blended into the drained coagulum to give the desired fat content, but otherwise the base was mixed to an even consistency, cooled and packaged as convenient. A starter culture dominated by *Leuconostoc* spp. and *Lact. lactis* biovar *diacetylactis* ensured that ymer had a pleasant flavour.

Modern systems of production involve concentrating the milk by a membrane process which removes the necessary volume of water – including some lactose and mineral salts – from the process mix prior to standardizing with cream and heat-treating at 80–85 °C for 5–10 minutes. Cooling to 22 °C is followed by inoculation and prolonged incubation for around 20 hours to the desired acidity. Gentle stirring is applied to give a smooth, homogeneous product that, on cooling, can be packaged for the retail market.

Lactofil is a somewhat similar product manufactured in Sweden, but a fat content of 5% is more usual.

Filmjolk

Filmjolk is an acidic Scandinavian milk with a fat content that ranges from 0.5% to 3%, depending on the market anticipated by the manufacturer. The starter culture is a mixture of mesophilic acid-producers and flavour-producers and, by employing an addition rate of up to 2% and a prolonged incubation time (18–20 hours), the final acidity of milk may be close to 1.0% lactic acid.

Nordic 'ropy' milks

Although mesophilic milks in general have a long tradition of consumption in Scandinavia, one particular variant that is thicker than normal milk and has a slightly slimy/ropy mouthfeel is taetmjolk (thick milk). Originally it was made by allowing the natural mesophilic flora to produce the desired acidity and flavour, while the addition of leaves of *Pinguicula vulgaris* or *Drosera* sp. provided a degree of thickening (Tamime and Robinson, 1988). These plants, known locally as 'thickening grass', are widespread in cold, damp northern latitudes, but the exact mode of their action does not appear to have been described. The end-product was a mildly acidic milk (pH 4.4) with a definite, pleasing consistency and, perhaps equally important for a rural community, a shelf-life of several weeks at low ambient temperature.

Nowadays, similar products can be manufactured by the judicious selection of starter cultures. The base – normal cow's milk with a fat content of around 3% – is severely heat-treated at 85–90 °C for 30 minutes, and then inoculated with a culture of *Leu. mesenteroides* subsp. *cremoris* together with a strain of *Lact. lactis* usually referred to as 'biovar *longi*'. This strain is unusual in that, in addition to secreting lactic acid, it synthesizes copious amounts of an extracellular protein. After incubation for 20–23 hours at 17–18 °C, the milk has acquired a mild acidity (pH 4.5–4.6) along with the ropy character of the traditional milk, and under controlled conditions (Alm, 1982).

Thermophilic lactic fermentations

In terms of world-wide popularity and volumes of production, fermented milks manufactured with strains of *Streptococcus salivarius* subsp. *thermophilus* (*Str. thermophilus*) and *Lactobacillus delbrueckii* subsp. *bulgaris* (*Lac. bulgaricus*) are dominant. As the name implies, the milks are normally fermented at a higher temperature than the mesophilic milks – usually around 42–43 °C. Due to the restricted range of organisms present, the products acquire a distinctive flavour and high acidity, with pH often as low as 3.8–4.0. The principal flavour compounds are aldehydes and ketones with acetaldehyde as the dominant component, and this limited spectrum not only serves to 'identify' the group of milks, but also ensures that the products have a 'clean' acidic flavour.

Yoghurt

Yoghurt is perhaps the most popular of all fermented milks, and its association with the thermophilic cultures is a reflection of its historical origins in the Middle East and/or India. By always storing milk in the same vessels, nomadic herders would have encouraged the build-up of a consistent microflora in the animal-skin bag or earthenware pot, so that as the milk soured it acquired a distinctive and (eventually) most acceptable flavour. The high acidity would have also ensured that any undesirable pathogenic bacteria were inactive, and hence it could have become a popular weaning food for young children. Obviously the microflora would have been uncontrolled in comparison with modern practices, but even the inevitable presence of yeasts could have proved beneficial as sources of B vitamins.

Modern production systems, such as that illustrated in Figure 1.21, depend on the availability of cultures of defined composition, and a manufacturer is able to choose with some degree of precision whether the end-product has a high level of flavour (acetaldehyde) or a more mild profile, and whether or not the texture is modified by the presence of extracellular polysaccharides synthesized by the bacteria. These latter decisions tend to be based upon the type of yoghurt to be manufactured, so that high flavour/low polysaccharide cultures may be selected for natural set yoghurts, and the reverse for stirred fruit brands.

Figure 9.1 Jars of concentrated yoghurt stored under olive oil, widely available in the Middle East; the product on the right is seasoned with thyme. (Courtesy of Dr A.Y. Tamime, West of Scotland College.)

Natural set yoghurt

As the name implies, these products have no added flavours, so that the perceived characteristics of the product depend entirely on metabolites secreted by the bacteria. Similarly the structure and texture of the yoghurt are derived solely from the protein gel formed by the acid coagulation of the milk – in most countries the use of stabilizers in set yoghurt is prohibited, and hence the products may well be much like the 'yoghurts' of biblical times.

As the milk component is critical for quality, the level of solids-not-fat (SNF) has to be raised to around 15–16% to ensure that sufficient protein is available to provide a firm gel. The addition of skim-milk powder to normal milk of 8.5–9.0% SNF is one available option, but concentration under vacuum or by ultra-filtration are often practised in larger dairies. Most of the yoghurt so manufactured is designated as 'low fat', which in practice means lower than normal levels of milk fat and hence anywhere between 0.5% and 2.0%; extra-low fat will be below 0.5%. However, although the fat does not influence the physical structure of the coagulum to any great extent, it does enhance the texture of the product as perceived by the consumer. For this reason, many consumers prefer the standard variety with around 3.5% fat and certainly the mouthfeel and satiety value of such products is vastly superior to the low-fat group.

At the other extreme, natural yoghurt is often used as an addition to a recipe and in this situation low cost is the primary objective. A typical example from this group, apart from the extra-low fat content, will have an SNF value of around 10–11%. In other words, sufficient protein is included to provide an acceptable gel, but the level is still low enough to allow the cost to be acceptable for culinary purposes.

The popular sizes in most parts of the world tend to be single portions of 150 g (125 g for form–fill–seal pots) with heat-sealed aluminium closures or 500 g with snap-on lids. The latter system allows product to be removed and the carton securely reclosed.

Stirred fruit yoghurt

Natural yoghurt is widely popular in the Middle East and the Mediterranean countries, as well as with consumers who enjoy it with fresh fruit or cereals. However, it has never enjoyed worldwide appeal. In the UK, for example, only 15% of total yoghurt production is the natural variety (Anon., 1992). The position is very different once fruit puree is added, and over 60% of all yoghurt sold in the UK is of the 'stirred fruit' type. The principal reason for this popularity is that the fruit and added sugar tend to mask the acidity of the product, so that whilst the pH may be as low as 3.8, the taste is not unpleasantly sharp. Consequently, fruit yoghurts have become pleasant lunch-time or supper snacks.

The basic production technique for the product is outlined in Figure 1.21, and the SNF is usually adjusted to around 13–14%. However, the composition has to reflect the fact that the yoghurt will be stirred after incubation, pumped to a unit for the addition of fruit and then pumped to a filling head for discharge into the cartons. Each of these steps takes its toll on the structure of the yoghurt, and hence most varieties have stabilizers added to ensure an attractive mouthfeel and consistency at the time of purchase. Starches and/or vegetable gums tend to be favoured by most manufacturers, with the balance between the alternatives being based upon the properties desired in the end-product. The choice of fruit is again a reflection of market demand, so that whilst ever-popular types like strawberry, raspberry and cherry appear under most labels, less popular fruits like gooseberry or unusual mixtures like peach and raspberry or apple and toffee are made available to satisfy a market niche or simply to offer the consumer an extended choice.

This desire to satisfy a demand from the market has resulted in a number of variations, and one of especial note is the 'thick and creamy' range of products. This range has levels of butterfat in the region of 4.5% and, for stirred fruit yoghurts, an above average content of non-fat milk solids and/or stabilizers. The end result is a yoghurt with definite 'luxury' appeal, in that the mouthfeel is rich and smooth compared with the standard product – so much so that many consumers ignore the additional calories and eat the yoghurt for pure pleasure. A rather different market niche is catered for by the 'organic yoghurts', but how many potential consumers are swayed by such claims is difficult to determine.

Labneh (Greek-style yoghurt)

It is widely accepted that natural yoghurt arose by the chance souring of milk, and that a desirable microflora evolved because farmers and households would have tended to use the same containers for storing milk. Flexible containers made of animal skins would have been the most convenient vessels for nomadic peoples, but one side-effect would have been that whey from the yoghurt would have seeped into the skin and been lost by evaporation. The end result would have been a yoghurt with a high total solids/semi-solid consistency and elevated level of acidity, and this concentrated yoghurt became known under a variety of local names.

Labneh is perhaps the most widely used name in the Middle East, but in Greece and elsewhere around the Mediterranean, the modern equivalent is known as 'strained yoghurt'. In essence, it is natural yoghurt that has been concentrated to around 22% total solids by ultra-filtration or centrifugation (Robinson and Tamime, 1993), and into which additional cream has been blended to give a final fat value of 10%. The acidity, which will have risen to 1.8–2.0% during the concentration step, is not excessive to the palate because the sharpness tends to be masked by the fat content. In addition, strained yoghurt tends to be eaten as a side-dish rather than as a 'snack', or forms the base for a 'dip' or special dish, such as the Greek favourite, tzatziki. This dish of strained yoghurt, cucumber and mint is an essential in any Greek restaurant, and typifies the current use of the product.

Skyr

Skyr is a fermented milk product that has for centuries enjoyed a unique place in the diet of Iceland. It is in some respects similar to quarg, and is widely consumed for breakfast or as a component of a light meal. According to legend, it was the early settlers from Norway who brought the product with them from their homeland, and whilst the product – although not the name – has vanished from the Nordic diet, its popularity in Iceland has remained undiminished.

Until some 50 years ago, the production of skyr was entirely farm-based. The first step was to remove the cream from the milk for other purposes. The skim-milk was then heated to 90–100 °C for a short period, prior to cooling to around 40 °C. Skyr from a previous batch, diluted with water for ease of dispersion into the skim-milk, provided the starter culture, and the inoculated milk was allowed to stand for 4–5 hours to coagulate; a low percentage of rennet was sometimes employed as well to assist with gel formation. This initial fermentation would have brought the pH of the milk down to around 4.6–4.7 and at this point the milk would have been cooled to around 20 °C. An overnight hold at this lower temperature would have allowed the pH to fall to around 4.2, and the coagulum was then suitable for concentration.

Linen bags provided the necessary vehicle; the curd was poured into the bag and was left for around 6 hours at 20 °C to drain. An overnight period of drainage at low temperature (>10 °C) raised the solids level into the region of 18–20%, and caused a further drop in pH to 3.8–4.0. Around 1 kg of skyr was derived from every 5 litres of skim-milk, and the quality varied greatly from farm to farm and region to region.

Nowadays, industrial production dominates the scene and the original process has been modified to meet the additional need for standardization and hygiene. After heat treatment for 30 minutes at 90 °C, the skim-milk is cooled to 40 °C and inoculated with a yoghurt culture consisting of *Str. thermophilus* and *Lac. bulgaricus*. Over the next 4–5 hours, the pH of the milk drops to 5.2, and the temperature is then reduced to 18–19 °C for the secondary, overnight incubation. Originally, it was assumed that this step was merely to further increase the acidity, but in fact it is the stage at which lactose-fermenting yeasts become active.

The end result is an acidic product (pH 4.1–4.2) with a distinctive flavour derived from the yeasts that are present as natural contaminants of the system, and up to 0.5% alcohol. Prolonged development of the yeasts could prove detrimental to the product and so the next step is pasteurization at just below 70 °C, i.e. sufficient heat to deplete the population of yeasts without excessive damage to the yoghurt microflora. Separation through a centrifugal separator follows to give a finished product with the following composition:

Total solids	17.5%
Protein	12.6%
Fat	0.2%
Lactose	3.8%
Minerals	0.8%
Titratable acidity	1.8% (as lactic acid)

By contrast, the traditional skyr had a solids content of over 20%, a protein level of 15.8% (Tamime and Robinson, 1988) and an acidity of 2.7%. Consequently, yields have improved to 1 kg of skyr/3.2 litres of milk, whilst the milder flavour has allowed the development of fruit-flavoured varieties which are consumed in the manner of snacks or desserts.

Shankleesh

Yoghurt and similar products are now sold almost universally through chill chains, and yet the original pressure to ferment milk came from the desire to preserve the material at ambient temperature. However, even yoghurt or its natural derivative, labneh, can soon become unpalatable due to over-acidification by the lactic acid bacteria and/or various biochemical reactions,

(a)

(b)

(c)

Figure 9.2 Shankleesh. (a) The concentrated yoghurt is hand-moulded into spheres around 5–6 cm in diameter. (b) Coating the spheres of concentrated yoghurt in thyme or other herbs. (c) This form of Shankleesh can be seen on display in supermarkets throughout the Middle East. (Courtesy of Dr I. Toufeili, American University of Beirut, Lebanon.)

and hence the evolution of products with longer storage lives was as logical progression.

Partial drying offered one solution. In many parts of the Middle East, labneh is moulded into spheres about the size of golf balls, and then surface-dried in the sun. The spheres are placed in glass jars (Figure 9.1) and covered with olive oil. In this state, the yoghurt will remain palatable for up to 12 months; when retrieved from the oil, it can be spread on bread in the manner of the original labneh.

The precise reasons for the excellence of this preservation process have yet to be elucidated, but it is perhaps the optimum procedure for retaining the natural properties of the material. It does, however, require access to suitable containers, and it may have been the scarcity of such receptacles that led to the arrival of shankleesh on the scene.

The basis for shankleesh is concentrated yoghurt, which is hand-moulded into large spheres (Figure 9.2(a)) and given a smooth outer surface. Coating with thyme or other herbs (Figure 9.2(b)) and immediate sale is one option (Figure 9.2(c)). Alternatively, the spheres may be placed in an earthenware storage jar and left at ambient temperature. Over the next few weeks, the spheres become colonized by a mixed flora of yeasts, such as *Debaryomyces hanseni*, and various species of *Penicillium*, of which *P. brevicompactum* is the most frequent isolate. The rapid development of this microflora not only gives the product a distinctive flavour, but may also reduce the risk of pathogenic organisms contaminating the product. The end result is that the original yoghurt has been provided with an extended shelf-life without recourse to refrigerated storage. Obviously, storage under olive oil has the advantage of retaining most of the properties of the natural yoghurt (labneh), but the production of shankleesh is preferred by a number of communities.

However, chemical deterioration and/or microbial spoilage remain a hazard, and it may be for this reason that some regions of the Middle East turned to sun-drying as the optimum solution, and produced kishk.

Kishk

The basis for kishk is again natural yoghurt, which may be concentrated prior to use and to which is added a quantity of wheat that has been dehusked, parboiled, dried and crushed. The end-product is a stiff, dough-like material that can be moulded into various shapes (the exact form depending on the region) before drying in the sun. After several days, the pieces of kishk will have become extremely dry and hard, and they can then be stored in open containers for two years or more at temperatures of 30 °C or above.

Although much of the nutritional value of the milk will be retained by this procedure, the organoleptic properties will be lost completely. Consequently, consumption involves grinding up the pieces of kishk in water, and then heating to make a gruel that is eaten rather in the manner of porridge. Clearly the end-product is not 'yoghurt', but at least the valuable components of the milk have been preserved for future use.

'Therapeutic' lactic fermentations

This group of products has many physical, chemical and organoleptic properties in common with the thermophilic milks described above, but the major difference lies in the cultures employed for the fermentation. The essential organisms are *Lactobacillus acidophilus* and various species of *Bifidobacterium*, including *Bif. bifidum*, *Bif. longum*, *Bif. adolescentis*, *Bif. breve* and *Bif. infantis*. The interest in these species lies in the fact they are all natural inhabitants of the human intestine.

The lactobacilli are dominant at the distal end of the small intestine, whilst the bifidobacteria are one of the major groups found in the colon. At their respective sites, the distinctive strains of these genera both occupy the lumen of the gut and colonize its walls. In these positions, the bacteria will:

- compete against undesirble bacteria, such as intestinal pathogens, for both nutrients and sites of attachment on cell walls;
- secrete lactic acid, and acetic acid as well in the case of *Bifidobacterium* spp., which will inhibit the growth of acid-intolerant bacteria such as *Salmonella* spp. or *Escherichia coli*; and
- secrete antibiotic compounds that will further enhance their competitive advantage.

The fact that these two genera are always present in high numbers in healthy adults is accepted as evidence that they are essential components of the intestinal microflora, and that if their numbers are depleted the effect is likely to be undesirable. In general, there is excellent evidence to support this view (Salminen and von Wright, 1993), and hence there is increasing acceptance of the idea that the regular ingestion of dairy products containing *Lac. acidophilus* and/or *Bifidobacterium* spp. may be beneficial to the average consumer (Sellars, 1991).

What has proved to be equally relevant is the fact that products manufactured with these cultures never develop the acid taste associated with normal yoghurt, for example, and consumers who have never featured fermented milks in their diet have now become converts. This reduced acidity is largely a reflection of the behaviour of the micro-organisms, for unlike some lactic acid bacteria, those species which were (prior to laboratory culture) inhabitants of the human intestine do grow well in milk and/or generate high levels of acid. The same species are also, in comparison with *Lac. bulgaricus* for example, intolerant of lactic acid, and hence the maintenance of products at mild acidities has the further advantage that it should allow the species to survive in a product throughout an anticipated shelf-life of 2–3 weeks.

Acidophilus milk

Although liquid milk containing lactic acid bacteria has no doubt been consumed for centuries, one of the first serious attempts to produce a 'health-promoting' fermented milk on a commercial scale involved Acidophilus milk in the USA.

The milk for this product can be skimmed or full-cream milk but, because *Lac. acidophilus* does not grow well in milk and would be easily overgrown by the usual microflora, the base milk has to be virtually 'sterile' at the time the culture is added. This need to eliminate competition means that the milk has to be severely heat-treated – either a double treatment at 90 °C with a rest period between the processes to allow any bacterial spores to germinate and new cells to be killed by the subsequent heating, or an Ultra-High Temperature procedure at 140 °C for 2 seconds. In either case, the 'clean' milk is then inoculated with high levels of *Lac. acidophilus* – up to 5% (v/v) with a liquid culture.

The milk is then left to incubate at 37 °C overnight, so that by the next morning the acidity of the product will have reached around 0.6–0.7% lactic acid. This slow rate of acid development (*vis-à-vis* yoghurt, for example) confirms the poor adaptation of *Lac. acidophilus* to milk but, as it happens, 0.6–0.7% lactic acid is about the level of tolerance of the culture anyway. Consequently, the optimum acidity is achieved by cooling the milk to 5 °C or less and halting any further activity by the culture; given time, the culture could generate up to 1.0–1.2% lactic acid, but the impact of such levels on cell viability over 2–3 weeks can be devastating in a low-solids product. After cooling, the Acidophilus milk is bottled and distributed under chilled conditions, and a shelf-life of about 2 weeks is to be expected.

The retail product enjoys a limited popularity mainly because, to many consumers, acidified milk is only acceptable as a 'thirst-quencher' on very hot days. Obviously, many people drink it for the beneficial influence of the culture, and in the Far East, for example, a similar product flavoured with fruit juice has proved to be extremely popular with children. Nevertheless, tastes in the Western world tend to veer away from acid, milk-based drinks, and the market niche for cultured drinks is now dominated by a variant, Sweet Acidophilis milk.

Sweet Acidophilis milk

This product is essentially standard pasteurized milk, either skimmed, semi-skimmed or full-cream, to which is added a sufficient volume of concentrated culture (frozen or freeze-dried) to give a viable cell count of *Lac. acidophilus* of 10×10^6/ml of product. The inoculated milk is then cooled and distributed in the manner of normal pasteurized. Because no fermentation has occurred, it tastes like the natural product and can therefore be consumed in a comparable manner to milk – except that it must not be heated. Yet it offers the health-promoting properties of the culture.

The level of addition is selected in the light of practical experience that has shown that detectable benefits for a consumer necessitate an intake, on a regular basis, of around 100–150 ml of a product containing at least one million viable cells of *Lac. acidophilus*. Given the natural taste of the product and its potential health-promoting properties, it is not surprising that several million litres of Sweet Acidophilus milk are drunk each year in the USA, along with an equal volume of a recent innovation – Bio-milk.

Bio-milk, or A/B (Acidophilus/Bifidus) milk, is manufactured in exactly the same manner as Sweet Acidophilus milk except that an equal quantity of a culture of *Bifidobacterium* sp. is added along with the culture of *Lac. acidophilus*. Both species are of human origin and offer potential benefits for the consumer, and the products are experiencing a growth in sales of some 30% per annum. How far the market will expand in North America and other warm climates like Australia remains to be seen, but it is of note that the European market tends to favour fermented products which can be eaten as snacks.

'Health-promoting' yoghurts

Although the word 'yoghurt' has in the past always been reserved for fermented milks containing *Lac. bulgaricus*, some countries are now allowing the term to be employed to describe a 'type of product', i.e. a cultured milk that has the physical characteristics usually associated with yoghurt, irrespective of the culture employed. An indication that the product is not standard yoghurt is usual, so that names like 'Bio-yoghurt', 'B/A yoghurt' or 'Mild yoghurt' tend to have found acceptance by the general public. The derivation of specific brand names has proved a less popular option, and hence most chill cabinets offer a range of 'health-promoting' products clearly identified as being broadly similar to yoghurt.

As with yoghurt, the products may be natural or flavoured set varieties, or stirred fruit variants, and are likely to be fermented with either *Lac. acidophilus* and *Bifidobacterium* sp. together, or in conjunction with *Str. thermophilus*. Both of the 'health-promoting' bacteria should be present at levels around 10×10^6 at the end of fermentation, as will be the culture of *Str. thermophilus*. The attraction to manufacturers of employing this latter organism is that, being well adapted to milk, it will rapidly metabolize some of the lactose in the milk to lactic acid, and so reduce the time needed to complete the fermentation. It is, in addition, a species that is no more acid tolerant than the principal cultures, and hence mild flavour and acid stability during distribution is still feasible.

The extent to which these 'health-promoting' milks will become established in the market-place remains to be seen, but it may be relevant that:

- in a number of European countries the bio-yoghurts are taking an appreciable and increasing share of the yoghurt market;
- health claims, such as 'the regular consumption of bio-milks/yoghurts can improve the digestive system', are allowed on the packaging for such products, and it seems likely that the range of claims will expand in the future (e.g. 'the regular consumption of bio-milks/yoghurts can help to alleviate the symptoms of lactose mal-digestion'); and
- clinical evidence in support of the health claims is beginning to accumulate, and hence reliance on anecdotal comments should soon be unnecessary.

Consequently, it would be fair to conclude that widespread acceptance of the theories relating to the therapeutic and/or prophylactic value of bio-products may not be far away, and that fermented milks will become a welcome component of many Western diets.

References

Abou-Donia, S.A. (1991) Manufacture of Egyptian, soft, pickled cheeses, in *Feta and Related Cheeses* (eds R.K. Robinson and A.Y. Tamime), Ellis Horwood, London.

Alm, L. (1982) *Report of the Department of Medical Nutrition*, Karolinska Institute, Stockholm.

Anifantakis, E.M. (1991) *Greek Cheeses*, National Dairy Committee of Greece, Athens.

Anon. (1992) The Eden Vale Review 1991. *Dairy Industries International*, 56(9), 23.

Atkinson, P.A. (1993) Mould-ripened Cheeses – II, in *Encyclopaedia of Food Science, Food Technology and Nutrition*, Volume 2 (eds R. Macrae, R.K. Robinson and M. Sadler), Academic Press Ltd, London.

Bottazzi, V. (1993) Manufacture of Extra-hard Cheeses, in *Encyclopaedia of Food Science, Food Technology and Nutrition*, Volume 2 (eds R. Macrae, R.K. Robinson and M. Sadler), Academic Press Ltd, London.

Chandan, R.C. (1991) Cheeses made by direct acidification, in *Feta and Related Cheeses* (eds. R.K. Robinson and A.Y. Tamime), Ellis Horwood, London.

Chapman, H.R. and Sharpe, M. Elisabeth (1990) Microbiology of Cheese, in *Dairy Microbiology*, Volume 2 (ed. R.K. Robinson), Elsevier Applied Science Publishers, London.

Cogan, T.M. and Accolas, J.-P. (1990) Starter cultures: types, metabolism and bacteriophage, in *Dairy Microbiology*, Volume 1 (ed. R.K. Robinson), Elsevier Applied Science Publishers, London.

Dalgleish, D.G., Horne, D.S. and Law, A.J.R. (1989) *Biochim. Biophys. Acta*, **991**, 383.

Davies, F.L. and Law, B.A. (eds) (1984) *Advances in the Microbiology and Biochemistry of Cheese and Fermented Milk*, Elsevier Applied Science Publishers, London.

Davis, J.G. (1976) *Cheese*, Volume III, Churchill Livingstone, London.

Fox, P.F. (ed.) (1989) *Developments in Dairy Chemistry*, **4**, Elsevier Applied Science Publishers, London.

Grandison, A.S. and Glover, F.A. (1993) Membrane Processing of Milk, in *Modern Dairy Technology*, Volume 1 (ed. R.K. Robinson), Elsevier Applied Science Publishers, London.

Guinee, T.P. and Fox, P.F. (1987) Salt in cheese: physical, chemical and biological aspects, in *Cheese*, Volume 1 (ed. P.F. Fox), Elsevier Applied Science Publishers, London.

Hofi, A.A., Youssef, E.H., Ghoneim, M.A. and Rawab, G.A. (1970) *Journal of Dairy Science*, 53(9), 1207.

IDF (1988) *Fermented milks – science and technology*, Bulletin No. 227, International Dairy Federation, B-1040 Brussels, Belgium.

Kosikowski, F.V. (1982) *Cheese and Fermented Milk Foods*, F.V. Kosikowski and Associates, New York.

Kurmann, J.A. and Rasic, J.Lj. (1991) The health potential of products containing bifidobacteria, in *Therapeutic Properties of Fermented Milks* (ed. R.K. Robinson), Elsevier Applied Science Publishers, London.

Law, B.A., Sharpe, M.E. and Chapman, H. (1976) *Journal of Dairy Research*, 43, 459.

Lawrence, R.C. and Gilles, J. (1987) Cheddar cheese and related dry-salted cheese varieties, in *Cheese*, Volume 2 (ed. P.F. Fox), Elsevier Applied Science Publishers, London.

Lenoir, J. and Tourneur, C. (1993) Mould-ripened Cheeses – I, in *Encyclopaedia of Food Science, Food Technology and Nutrition*, Volume 2 (eds R. Macrae, R.K. Robinson and M. Sadler), Academic Press Ltd., London.

Marcos, A. (1987) Spanish and Portuguese Cheese Varieties, in *Cheese*, Volume 2 (ed. P.F.

Fox), Elsevier Applied Science Publishers, London.

Marshall, Valerie M.E. (1987) *Journal of Dairy Research*, **54**, 559.

Papageorgiou, D.K. and Marth, E.H. (1989) *Journal of Food Protection*, **52**(2), 82.

Prato, S. del (1993) Mozzarella Cheese. *Dairy Industries International*, **58**(4), 26.

Robinson, R.K. (1988) Cultures for Yoghurt – their selection and use. *Dairy Industries International*, **53**(7), 15.

Robinson, R.K. (1991) Micro-organism of Fermented Milks, in *Therapeutic Properties of Fermented Milks* (ed. R.K. Robinson), Elsevier Applied Scientific Publishers, London.

Robinson, R.K. and Samona, Aspasia (1992) Health aspects of 'bifidus' products. *International J. Food Sciences and Nutrition*, **43**, 175.

Robinson, R.K. and Tamime, A.Y. (1990) Microbiology of Fermented Milks, in *Dairy Microbiology*, Volume 2 (ed. R.K. Robinson), Elsevier Applied Science Publishers, London.

Robinson, R.K. and Tamime, A.Y. (1993) Manufacture of Yoghurt and other Fermented Milks, in *Modern Dairy Technology*, Volume 2 (ed. R.K. Robinson), Elsevier Applied Science Publishers, London.

Salminen, S. and von Wright, A. (eds) (1993) *Lactic Acid Bacteria*, Marcel Dekker Inc., New York.

Scott, R. (1986) *Cheesemaking Practice*, 2nd edn, Elsevier Applied Science Publishers, London.

Sellars, R.L. (1991) Acidophilus Products, in *Therapeutic Properties of Fermented Milks* (ed. R.K. Robinson), Elsevier Applied Science Publishers, London.

Seth, Rachel J. and Robinson, R.K. (1988) Factors contributing to the flavour characteristics of mould-ripened cheese, in *Developments in Food Microbiology*, 4 (ed. R.K. Robinson), Elsevier Applied Science Publishers, London.

Shawn, M.B. (1993) Modern cheesemaking – soft cheeses, in *Modern Dairy Technology* (ed. R.K. Robinson), Chapman & Hall, London.

Stobbs, W. (1984) *Guide to Cheeses of France*, The Apple Press Ltd, London.

Tamime, A.Y. (1993) Modern Cheesemaking: Hard Cheeses, in *Modern Dairy Technology*, Volume 2 (ed. R.K. Robinson), Elsevier Applied Science Publishers, London.

Tamime, A.Y. and Deeth, H.C. (1980) *Journal of Food Protection*, **43**, 939.

Tamime, A.Y. and Kirkegaard, J. (1991) Manufacture of Feta cheese – industrial, in *Feta and Related Cheeses* (eds R.K. Robinson and A.Y. Tamime), Ellis Horwood, London.

Tamime, A.Y. and Robinson, R.K. (1985) *Yoghurt – science and technology*, Pergamon Press, Oxford.

Tamime, A.Y. and Robinson, R.K. (1988) *Journal of Dairy Research*, **55**, 281.

Tamime, A.Y., Dalgleish, D.G. and Banks, W. (1991) Introduction, in *Feta and Related Cheeses* (eds R.K. Robinson and A.Y. Tamime), Ellis Horwood, London.

USDA (1974) *Cheese Varieties and Descriptions*. US Department of Agriculture, Handbook No. 54.

Zerfiridis, G. and Pappas, C. (1989) *Technology of Greek Cheeses*, Proceedings of the Greek National Dairy Committee, Athens.

Further reading

Fox, P.F. (ed.) (1987) *Cheese*, Volumes 1 and 2, Elsevier Applied Science Publishers, London.

Hui, Y.H. (ed.) (1992) *Dairy Science and Technology Handbook*, Volumes 1–3, VCH Publishers Inc., New York.

Marshall, Valerie M.E. (1984) Flavour development in fermented milks, in *Advances in the Microbiology and Biochemistry of Cheese and Fermented Milk* (eds F.L. Davies and B.A. Law), Elsevier Applied Science Publishers, London.

Index

Page numbers appearing in **bold** refer to figures and page numbers appearing in *italic* refer to tables.